Optimization of Water-Level Monitoring Networks in the Eastern Snake River Plain Aquifer Using a Kriging-Based Genetic Algorithm Method

By Jason C. Fisher

DOE/ID-22224
Prepared in cooperation with the Bureau of Reclamation and
U.S. Department of Energy

Scientific Investigations Report 2013–5120

U.S. Department of the Interior
U.S. Geological Survey

U.S. Department of the Interior
SALLY JEWELL, Secretary

U.S. Geological Survey
Suzette M. Kimball, Acting Director

U.S. Geological Survey, Reston, Virginia: 2013

For more information on the USGS—the Federal source for science about the Earth, its natural and living resources, natural hazards, and the environment, visit http://www.usgs.gov or call 1–888–ASK–USGS.

For an overview of USGS information products, including maps, imagery, and publications, visit http://www.usgs.gov/pubprod

To order this and other USGS information products, visit http://store.usgs.gov

Suggested citation:
Fisher, J.C., 2013, Optimization of water-level monitoring networks in the eastern Snake River Plain aquifer using a kriging-based genetic algorithm method: U.S. Geological Survey Scientific Investigations Report 2013-5120 (DOE/ID-22224), 74 p., http://pubs.usgs.gov/sir/2013/5120/.

Contents

Figures

Tables

Conversion Factors, Datums, and Abbreviations and Acronyms

Conversion Factors

SI to Inch/Pound

Multiply	By	To obtain
Length		
centimeter (cm)	0.3937	inch (in.)
meter (m)	3.281	foot (ft)
kilometer (km)	0.6214	mile (mi)
meter (m)	1.094	yard (yd)
Area		
square meter (m^2)	0.0002471	acre
square kilometer (km^2)	247.1	acre
square meter (m^2)	10.76	square foot (ft^2)
square kilometer (km^2)	0.3861	square mile (mi^2)
meter (m)	1.094	yard (yd)
Slope		
meter per kilometer (m/km)	5.28	foot per mile (ft/mi)

Conversion Factors, Datums, and Abbreviations and Acronyms—Continued

Datums

Vertical coordinate information is referenced to the North American Vertical Datum of 1988 (NAVD 88).

Vertical coordinate information for historical data collected and stored as National Geodetic Vertical Datum of 1929 (NGVD 29) has been converted to the commonly used NAVD 88 for this publication. Conversion between NAVD 88 and NGVD 29 varies spatially, and is accurate to within about plus or minus 2 centimeters (cm). The reader is directed to the National Geodetic Survey website for VERTCON at http://www.ngs.noaa.gov/TOOLS/Vertcon/vertcon.html for a detailed description of the height conversion methodology.

Elevation, as used in this report, refers to distance above the vertical datum.

Horizontal coordinate information is referenced to the North American Datum of 1983 (NAD 83).

Conversion of horizontal coordinate information from North American Datum of 1927 (NAD 27) to NAD 83 varies spatially, and typically is accurate to within 12–18 cm. For assistance with conversions, the reader is directed to the National Geodetic Survey website for NADCON at http://www.ngs.noaa.gov/TOOLS/Nadcon/Nadcon.html.

Maps are based on the Albers Equal Area Conic projection using a central meridian of 113° W., standard parallels of 42° 50′ N. and 44° 10′ N., a false easting of 200,000.00 meters, and the latitude of the projection's origin at 41° 30′ N.

Abbreviations and Acronyms

Abbreviation or acronym	Definition
cov	covariance
c	sill of the theoretical semivariogram
e	residual component
f	individual objective function
g	nugget of the theoretical semivariogram
h	distance between two measurement points
$\tilde{h}$	lag distance
ℓ	length of a square block side in the uniform kriging grid
m	trend function
n	number of measurement sites in the monitoring network
n_a	maximum number of times a child chromosome can be aborted during crossover
n_e	number of measurement sites in the existing monitoring network
n_{iter}	maximum number of iterations in the genetic algorithm
n_n	number of nodes in the kriging grid
n_{pen}	number of times the penalty function is invoked
n_{pop}	number of chromosomes in the population
n_r	number of observation wells to remove from an existing monitoring network
n_{run}	maximum number of consecutive iterations without any improvement in the best fitness value
n_v	number of points in the empirical semivariogram
p-value	probability of an observed result arising by chance

Conversion Factors, Datums, and Abbreviations and Acronyms—Continued

Abbreviations and Acronyms

Abbreviation or acronym	Definition
r	range of the theoretical semivariogram
s	Cartesian coordinate at a geographic location
s'	Cartesian coordinate at a geographic location
s_0	estimation point
s_n	Cartesian coordinate at a node in the kriging grid
var	variance
w	weighting coefficient
x	easterly coordinate
$\boldsymbol{x}$	decision variables
y	northerly coordinate
z	water-level elevation
$\hat{z}$	estimated water-level elevation
$\hat{z}_{orig}$	estimated water-level elevation based on measurements from the original monitoring network
z^*	estimated water-level elevation from leave-one-out cross validation
ACO	ant colony optimization
BB	branch-and-bound
C	penalty coefficient
C_e	covariance function of the residual component
Co-op network	Federal-State Cooperative water-level monitoring network
CRAN	Comprehensive R Archive Network
DOE	U.S. Department of Energy
E	expected value
ESRP	eastern Snake River Plain
F	weighted sum objective function
F'	fitness function
GA	genetic algorithm
GB	gigabytes
GDAL	Geospatial Data Abstraction Library
GTS	geostatistical temporal/spatial
IDW	inverse distance weighting
IDWR	Idaho Department of Water Resources
INL	Idaho National Laboratory
KED	kriging with external drift
MAROS	monitoring and remediation optimization system
N_h	number of data pairs in a bin
OK	ordinary kriging
P	penalty function

Conversion Factors, Datums, and Abbreviations and Acronyms—Continued

Abbreviations and Acronyms

Abbreviation or acronym	Definition
PLE	percent local error
PROJ.4	cartographic projections library
R^2	coefficient of determination
R	set of all real numbers
$\bar{R}^2$	adjusted coefficient of determination
RAM	random-access memory
Reclamation	Bureau of Reclamation
RMSD	root-mean-square deviation
RMSE	root-mean-square error
SA	simulated annealing
SVMs	support vector machines
UK	universal kriging
USGS	U.S. Geological Survey
USGS-INL	U.S. Geological Survey-Idaho National Laboratory
USGS-INL network	U.S. Geological Survey-Idaho National Laboratory water-level monitoring network
Z	set of all integers
X	random variable
Y	random variable
β	deterministic trend coefficient
γ_e	semivariogram of the residual component
$\widehat{\gamma}_e$	empirical semivariogram of the residual component
$\tilde{\gamma}_e$	theoretical semivariogram of the residual component
Δh	constant bin width
ε_z	measurement error for the water-level elevation
λ	kriging coefficient
μ_e	mean of the residual component
σ_e^2	variance of the residual component
σ_{UK}	standard error from universal kriging
σ_{UK}^2	estimated variance from universal kriging
σ_z	standard deviation of water-level measurements

Optimization of Water-Level Monitoring Networks in the Eastern Snake River Plain Aquifer Using a Kriging-Based Genetic Algorithm Method

By Jason C. Fisher

Abstract

Long-term groundwater monitoring networks can provide essential information for the planning and management of water resources. Budget constraints in water resource management agencies often mean a reduction in the number of observation wells included in a monitoring network. A network design tool, distributed as an R package, was developed to determine which wells to exclude from a monitoring network because they add little or no beneficial information. A kriging-based genetic algorithm method was used to optimize the monitoring network. The algorithm was used to find the set of wells whose removal leads to the smallest increase in the weighted sum of the (1) mean standard error at all nodes in the kriging grid where the water table is estimated, (2) root-mean-squared-error between the measured and estimated water-level elevation at the removed sites, (3) mean standard deviation of measurements across time at the removed sites, and (4) mean measurement error of wells in the reduced network. The solution to the optimization problem (the best wells to retain in the monitoring network) depends on the total number of wells removed; this number is a management decision. The network design tool was applied to optimize two observation well networks monitoring the water table of the eastern Snake River Plain aquifer, Idaho; these networks include the 2008 Federal-State Cooperative water-level monitoring network (Co-op network) with 166 observation wells, and the 2008 U.S. Geological Survey-Idaho National Laboratory water-level monitoring network (USGS-INL network) with 171 wells. Each water-level monitoring network was optimized five times: by removing (1) 10, (2) 20, (3) 40, (4) 60, and (5) 80 observation wells from the original network. An examination of the trade-offs associated with changes in the number of wells to remove indicates that 20 wells can be removed from the Co-op network with a relatively small degradation of the estimated water table map, and 40 wells can be removed from the USGS-INL network before the water table map degradation accelerates. The optimal network designs indicate the robustness of the network design tool. Observation wells were removed from high well-density areas of the network while retaining the spatial pattern of the existing water-table map.

Introduction

Long-term groundwater monitoring networks have provided vital information for sustainable water resources management in the eastern Snake River Plain (ESRP) (figs. 1 and 2), Idaho. Data from these networks have been used to validate groundwater flow models, to evaluate the response of groundwater levels to artificial recharge efforts and changing climatic drivers, and to review water rights with respect to the long-term sustainability of aquifer resources. Given the high costs associated with the maintenance of these networks, development of efficient network designs is essential. The design of a groundwater-level monitoring network is dependent on the spatial and temporal distribution of water levels in the aquifer. These distributions are extremely complicated in the ESRP given its diverse geology, perched alluvial conditions that overlie the regional aquifer, and variable fluxes between groundwater and surface water (Whitehead, 1992). Care must be taken to include the many complex factors involved when describing the groundwater system; for example, statistical procedures must be used to simulate the water-table surface (that is, the surface where the water pressure head is equal to the atmospheric pressure) of the aquifer. A more efficient network of monitoring wells may be established by evaluating the value of observations measured at each well and the degree to which observations are redundant, and then removing low-value or redundant wells from the observation network. A heuristic optimization procedure for identifying these redundant wells is presented in this report. Heuristic is a technique for efficiently guiding the process of optimization; it does not guarantee that the best solution will be found. This study was conducted by the U.S. Geological Survey (USGS) in cooperation with the U.S. Bureau of Reclamation (Reclamation) and the U.S. Department of Energy (DOE).

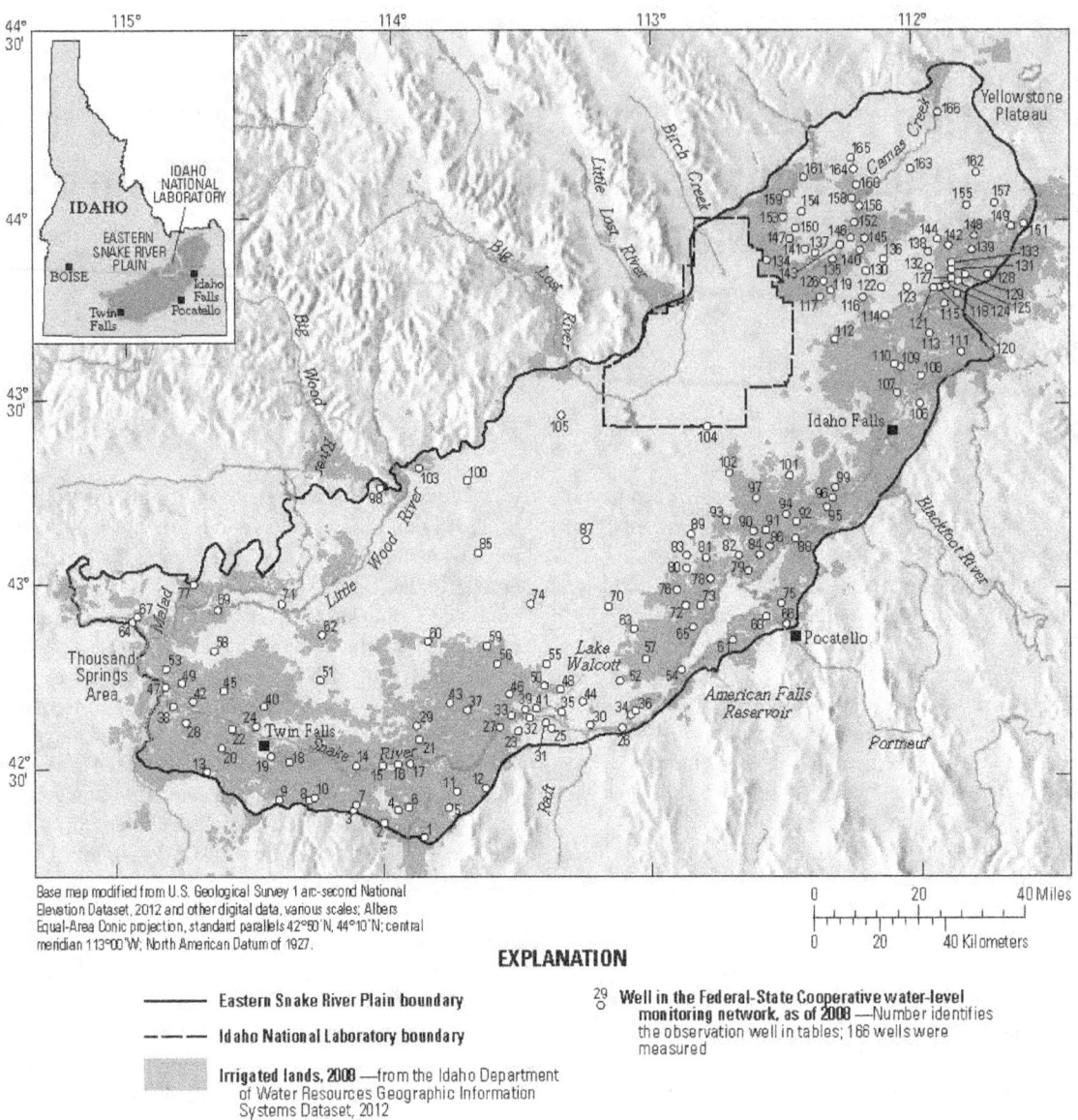

Figure 1.   Locations of wells in the Federal-State Cooperative water-level monitoring network, eastern Snake River Plain, Idaho, 2008.

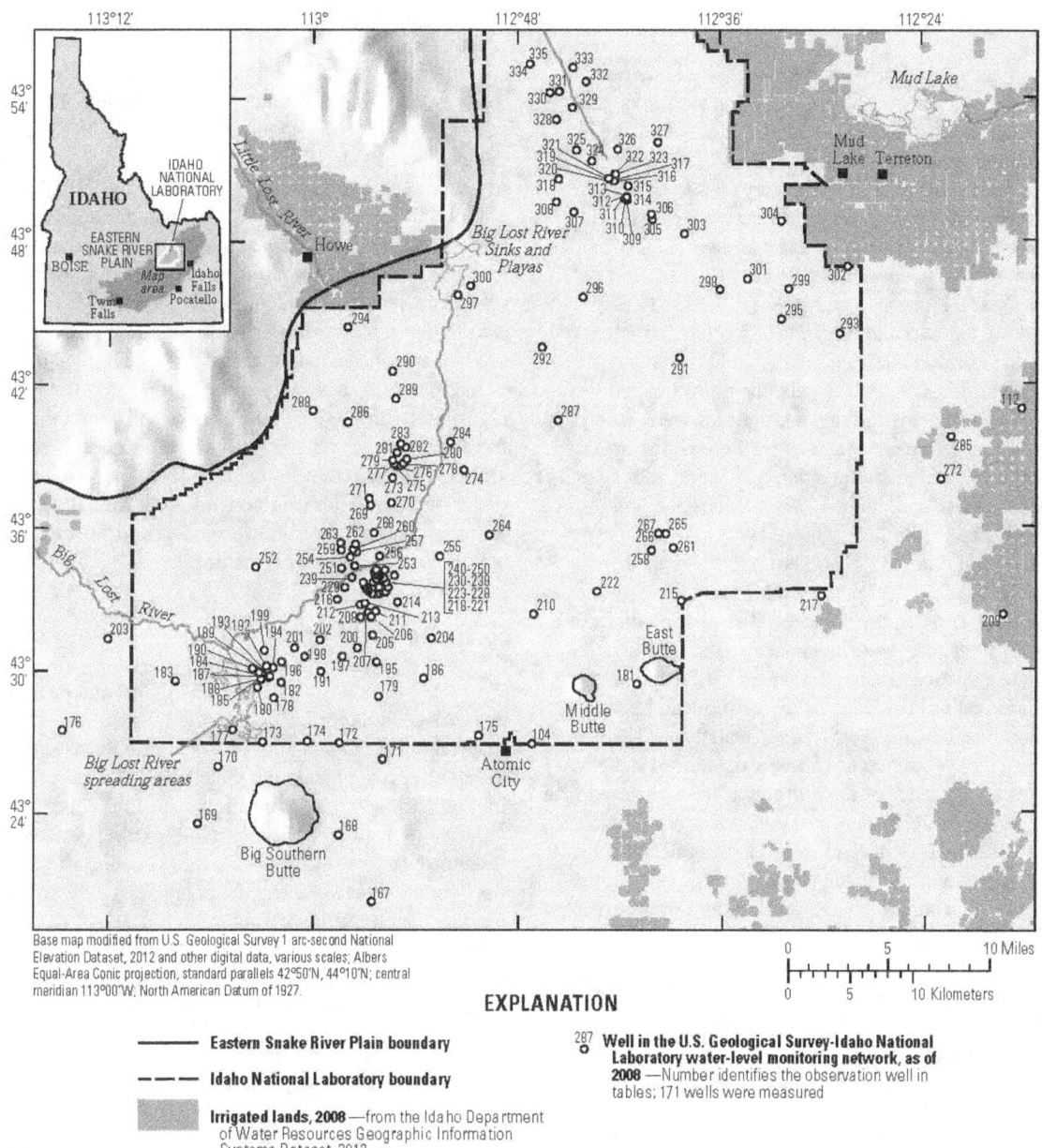

EXPLANATION

——— **Eastern Snake River Plain boundary**

– – – **Idaho National Laboratory boundary**

Irrigated lands, 2008 —from the Idaho Department
of Water Resources Geographic Information
Systems Dataset, 2012

$\frac{287}{\circ}$ **Well in the U.S. Geological Survey-Idaho National
Laboratory water-level monitoring network, as of
2008** —Number identifies the observation well in
tables; 171 wells were measured

Figure 2.   Locations of wells in the U.S. Geological Survey-Idaho National Laboratory water-level monitoring
network, Idaho National Laboratory and vicinity, Idaho, 2008.

Previous Investigations

Groundwater monitoring networks can be classified into two categories: (1) water-quality monitoring networks, and (2) water-level monitoring networks. A typical objective for long-term monitoring of water quality is the development of a cost-effective design that adequately characterizes a contaminant plume. Examples of such networks are given by Grabow and others (1993), Reed and others (2000), Lin and Rouhani (2001), Cameron and Hunter (2002), Aziz and others (2003), Passarella and others (2003), Nunes and others (2004b), Herrera and Pinder (2005), Yeh and others (2006), Li and Hilton (2007), and Dhar and Datta (2010). For long-term monitoring of water levels, the typical objective is the development of a cost-effective design for water resources management that retains adequate overall prediction accuracy. Examples of such networks are given by Gangopadhyay and others (2001), Asefa and others (2004), Nunes and others (2004a), and Khan and others (2008).

Loaiciga and others (1992) classified the general approaches to network design into two categories: (1) hydrogeologic, when no advanced statistical methods are used; and (2) statistical, when advanced statistical methods are used. The statistical approach to network design can include numerical models of groundwater flow and transport, interpolation methods, and (or) statistical analysis. The variance-based (also known as variance reduction) statistical method uses the variance of the estimation error obtained from kriging to assess the suitability of a given network (Nunes and others, 2004a, 2004b). A given monitoring network has an uncertainty (quantified as the variance of the estimation error) that depends on the number and position of observation wells in the monitoring network. Adding wells to this network usually decreases uncertainty, whereas removing wells usually increases it. These methods systematically search for a set (that is, number and position) of observation wells that minimize the variance of the estimation error.

In previous investigations, heuristic search methods, including decision support tools and mathematical optimization, were used to identify the best set of observation wells in a monitoring network. Examples of decision support tools include the monitoring and remediation optimization system (MAROS) (Aziz and others, 2003) and the geostatistical temporal/spatial (GTS) algorithm (Cameron and Hunter, 2002). Decision support tools typically are applicable only to a specific class of problems; therefore, mathematical optimization techniques that are applicable to a variety of problems have been adopted much more widely for network design problems. Reducing the number of observation wells in an existing monitoring network is a non-linear combinatorial problem and, therefore, is well suited for heuristic algorithms. Genetic algorithms (GAs), simulated annealing (SA), support vector machines (SVMs), and colony optimization (ACO), and the branch-and-bound (BB) algorithm are search heuristics that have been used to optimize long-term-monitoring networks. For example, Reed and others (2000) optimized monitoring networks using inverse distance weighting (IDW) and ordinary kriging (OK) using heuristic GAs and simulation models. Similarly, Nunes and others (2004a) used SA with statistical methods to maximize spatial accuracy and to minimize temporal redundancy of a large groundwater monitoring network. Asefa and others (2004) present a methodology based on SVMs for designing a water-level monitoring network that identifies observation well locations based on their importance in explaining the potentiometric surface. Li and Hilton (2007) used an ACO algorithm with IDW to derive a reduced configuration of a trichloroethylene concentration monitoring well network. Dhar and Datta (2010) developed a methodology for designing a water-quality monitoring network by formulating the problem as a logic-based mixed-integer linear optimization model and solving it using the BB algorithm.

Purpose and Scope

The purpose of this report is to present a heuristic network design tool for optimizing long-term monitoring networks. This tool is applied to determine cost-effective designs for two preexisting water-level monitoring networks in the ESRP aquifer. Monitoring costs are reduced by eliminating data redundancy in the existing networks—that is, by removing observation wells that have little-to-no effect on the overall water table characterization. A genetic algorithm is used to search for the optimal network design using multiple objectives to evaluate candidate subsets of observation wells within an existing monitoring network. The design objectives considered are to: (1) minimize the interpolation error in the water-table map, (2) preserve local anomalies in the water-table surface, (3) preserve the variability of water-level measurements over time, and (4) maintain wells with higher measurement accuracy. The individual objectives are combined into a single composite objective function. The spatial interpolation technique, universal kriging (UK), is embedded in the optimization formulation for estimating water-level elevations at unmonitored locations. The total number of wells to remove from the original network is a management decision. Network efficiency is expected to change as more data and resources become available; therefore, a reexamination of the network design every few years may help determine the effectiveness of a groundwater monitoring program. The network design tool is applied separately to each water-level monitoring network in the ESRP.

The model of the semivariogram, which describes the spatial variability of the water table and is used as input for UK, is calculated using 2008 median water-level measurements in observation wells of both monitoring networks. Because of budgetary constraints on the groundwater monitoring programs shortly after 2008, this period was selected. The optimized monitoring networks provided in this report can be used to evaluate observation well reductions made since 2008 through a comparison of the optimized networks with the networks currently (2013) in use, and to facilitate the design of future groundwater monitoring networks in the ESRP aquifer.

Description of Study Area

The study area is the ESRP in Idaho, a relatively flat topographic depression, about 270 km long and 100 km wide (fig. 1). Land surface elevations range from about 700 m above the North American Vertical Datum of 1988 (NAVD 88) near the Thousand Springs area in the southwest to more than 2,000 m near the Yellowstone Plateau in the northeastern part of the plain. The ESRP crosses the roughly northwest fault-bounded mountain ranges of eastern Idaho from southwest to northeast (fig. 1). The steep mountain ranges bordering the plain are about 2,000–3,500 m in elevation, and collect as much as 150 cm per year of snow during the winter, which melts in late spring and early summer. The ESRP is a semiarid sagebrush steppe with warm summers and cold winters, and precipitation averaging 20 cm per year (Garabedian, 1992, p. 2).

The ESRP is the track of the time-transgressive Yellowstone Hotspot super-caldera eruptions (Pierce and Morgan, 1992). Each caldera system began as a high geoid anomaly, followed by several super-caldera eruptions of mostly rhyolitic material. After each super-caldera eruptions ceased, a quiescent period followed, during which basalt erupted in response to continued high heat flow. Post super-caldera, thermal contraction, and emplacement of a mafic sill in the mid-crust caused continuing subsidence (Rodgers and others, 2002).

The entire ESRP is subsiding, although subsidence is not uniform and localized depocenters collect sediment and are thought to have considerable control on groundwater movement in the aquifer (Fisher and Twining, 2011, p. 36). The Idaho National Laboratory (INL) occupies a prominent depocenter, informally named the Big Lost Trough, which is bounded on the north, east, and south by volcanic highlands. The Big Lost Trough may have as much as 15 percent sediment interbedded between basalt flows (Anderson and Liszewksi, 1997). Sediment also accumulates along the margins of the ESRP and is typically accompanied by agriculture in these areas (fig. 1).

The ESRP aquifer generally is considered an unconfined aquifer; however, sediment layers near the INL behave as confining units (Fisher and Twining, 2011, p. 34). Groundwater flows in a regionally southwest direction and discharges mainly through large springs and seeps along the Snake River in the Thousand Springs area in the southwestern part of the plain (fig. 1). Groundwater moves horizontally through basalt interflow zones, and vertically through joints and interfingering edges of interflow zones. Infiltration of surface water, heavy pumpage, geologic conditions, and seasonal fluxes of recharge and discharge locally affect the movement of groundwater in the aquifer. Recharge to the aquifer is from infiltration of precipitation, groundwater inflow from tributary drainages, infiltration of surface water diverted for irrigation, and stream and canal losses (Garabedian, 1992, p. 11). Land irrigated with groundwater on the ESRP is along the southeastern and southern margins of the plain, from north of Idaho Falls to west of Twin Falls, and in the Mud Lake area northeast of the INL (figs. 1 and 2) (Ackerman and others, 2006, p. 6).

Water-Level Monitoring Networks

Long-term water-level monitoring networks were established in the ESRP aquifer to identify changes in storage and the general rate and direction of groundwater flow in the aquifer. In this study, only water levels representative of the water table in the ESRP aquifer were important. For example, observation wells screened across or just below the water table are excellent indicators of the water-table elevation, whereas, wells screened in locations of perched groundwater or deeper confined aquifers are poor indicators and were excluded from the analysis of the monitoring network.

Federal-State Cooperative Water-Level Monitoring Network

The Federal-State Cooperative water-level monitoring network (Co-op network) is administered by the USGS, Reclamation, and the Idaho Department of Water Resources (IDWR). At the end of 2008, water-level elevations in 166 observation wells were measured annually, semi-annually, quarterly, bi-monthly, or monthly by the USGS and Reclamation. The spatial distribution of these wells in the ESRP is shown in figure 1. Observation wells cover most of the plain, except for the INL. Site information (such as, local name, map number, and site number) for each well in the network is given in table 5 (at back of report).

U.S. Geological Survey-Idaho National Laboratory Water-Level Monitoring Network

The U.S. Geological Survey-Idaho National Laboratory (USGS-INL) water-level monitoring network (USGS-INL network) is administered by the USGS-INL Project Office in cooperation with the DOE. At the end of 2008, water-level elevations in 171 observation wells were measured annually, semi-annually, quarterly, or monthly by the USGS. The spatial distribution of these wells in and near the INL is shown in figure 2. The west, east, north, and south bounding coordinates of these wells are about 113°17', 112°17', 43°57', and 43°19', respectively. USGS-INL network coverage is densest around INL facilities. Site information for each well in the network is given in table 5.

Methods

Sources and Descriptions of Data

Existing sources of information for the observation wells include: geographic coordinates (that is, longitude, latitude, and elevation) of the land surface reference point for water-level measurements (measurement point); and depth-to-water measurements. Depth-to-water measurements are easily converted to water-level elevations by subtracting depth-to-water measurements from the elevation at land surface.

Geographic Coordinates

Methods for determining the spatial location of the land surface reference point of a well varied throughout the networks. The least accurate geographic coordinates were interpolated from USGS topographic maps and were accurate to about plus-or-minus (±) 30 m (or 1 arc-second) in the horizontal direction and to about ± 3.66 m in the vertical direction. The most accurate coordinate positions were determined by a professional land surveyor licensed in the State of Idaho using a Differential Global Positioning System. Surveyed positions were accurate to about ± 0.3 m (0.01 arc-second) in the horizontal direction and ± 0.003 m in the vertical direction. The horizontal position is expressed in latitude and longitude in conformance with the North American Datum of 1983 (NAD 83). The vertical position is

expressed as the elevation above the NAVD 88. Geographic coordinates of the land surface reference point at each observation well are given in table 5. For wells in both monitoring networks (the number of sites [n] = 335), the mean reference-point error is 0.36 m, with a standard deviation of 0.78 m.

Water Levels

Water levels were obtained by subtracting the depth to water from the elevation of the land-surface measurement point. Water-depth measurements were obtained by the USGS using steel or electric measuring tapes. The depth to the water level below the land-surface reference point were accurate to ± 0.01 m (± 0.02 ft), although a few measurements obtained under less-than-ideal conditions (for example, when condensation accumulated in the borehole) had errors greater than ± 0.01 m. The period of record and frequency of monitoring is variable for each observation well, with the earliest water-levels recorded in 1922. The historical variability of water-level measurements in a well over the entire period of record is described with the standard deviation (σ_z) (table 5). The standard deviation is a measure of the seasonal fluctuations in the water table as well as long-term trends. Standard deviations are shown spatially and proportionally in figure 3. For wells in both monitoring networks, the standard deviation ranged from 0.15 to 10.31 m, with a median value of 1.55 m. These standard deviations are small relative to the range of measured hydraulic heads across the region.

Water-level data obtained during the 2008 calendar year were used to estimate the water-table surfaces. In 2008, the number of measurements collected in each well ranged from 1 to 63, with a mean of 5. The 2008 median water-level elevation was determined for each well and expressed as an elevation above the NAVD 88 (table 5). Summing the measurement errors of measurement point elevation and depth-to-water gives the measurement error of the water-level elevations. This error estimate assumes that the borehole is vertical. The mean measurement error for 2008 water-level elevations was determined for each observation well (table 5, fig. 4) and ranged from 0.01 to 3.66 m, with a median value of 0.01 m. Mean measurement errors greater than 0.03 m in certain observation wells were because of the large measurement error for the elevation of the land surface reference point at these wells.

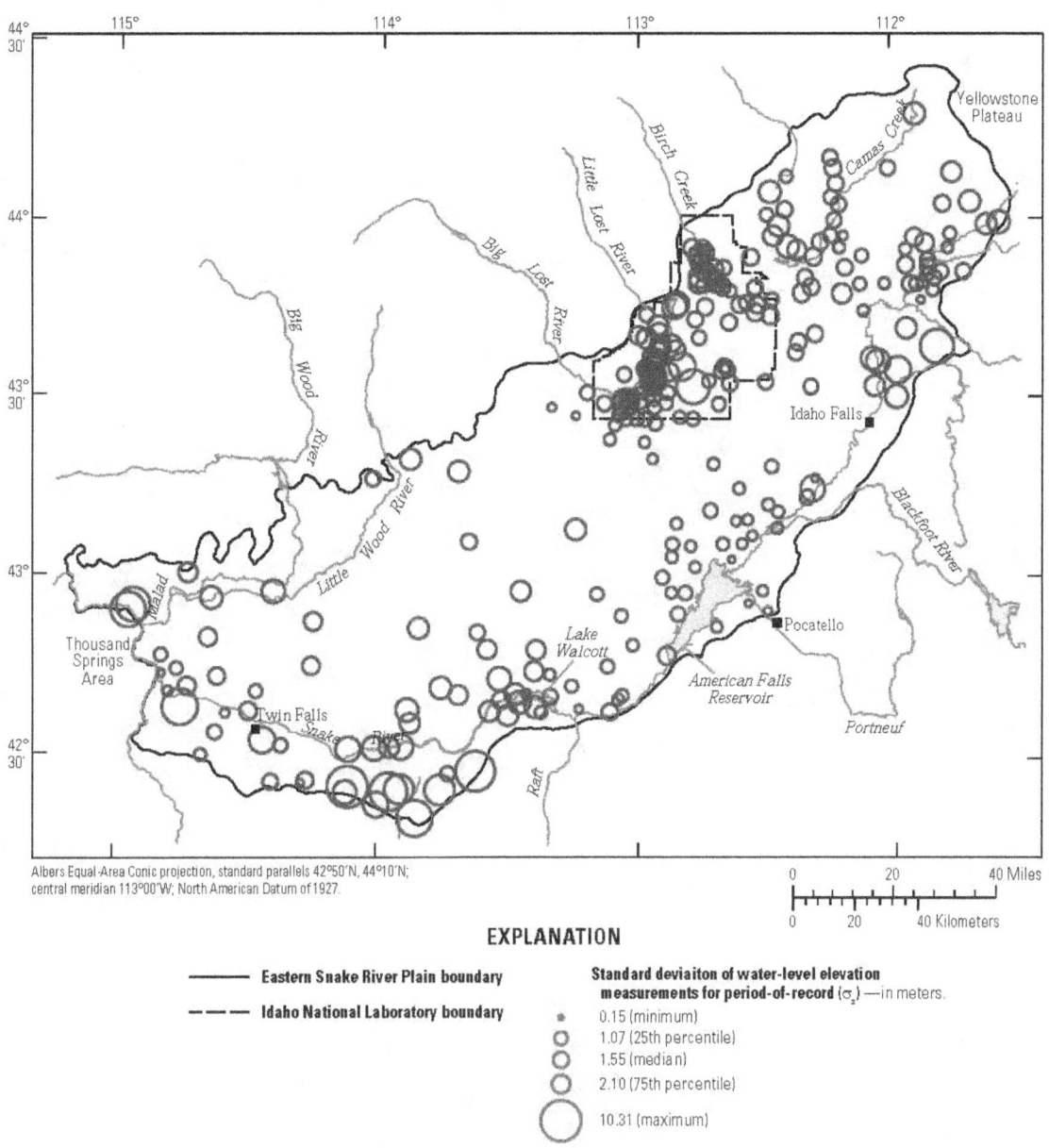

Figure 3. Standard deviation of water-level measurements for the entire period-of-record (duration varies at each well site), eastern Snake River Plain, Idaho.

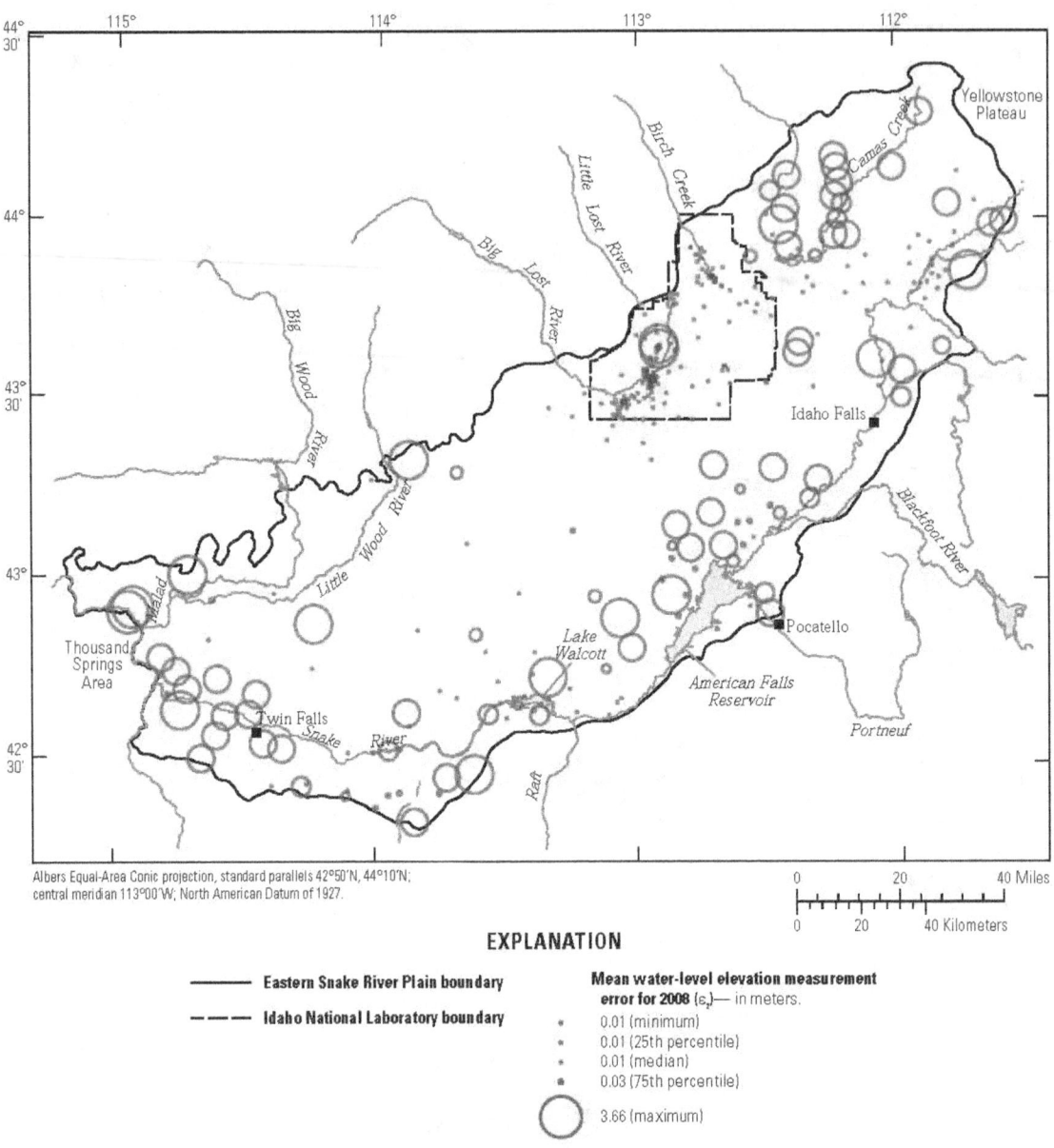

Figure 4.   Mean measurement error of water-level elevations, eastern Snake River Plain, Idaho, calendar year 2008.

Interpolation of the Water Table

The geostatistical technique known as kriging is used to interpolate and extrapolate the water-level elevation at unmeasured locations in the ESRP aquifer. Snyder (2008, p. 19) describes kriging as a type of spatial moving average, where the value at an unmeasured location is estimated as a weighted average of the measured values. The weights assigned to the measured values depend on spatial trends and possible correlations in the data (Bossong and others, 1999, p. 4). Correlation between measurements at two sites is assumed to depend on the separation distance between the two sites. Generally, measurement sites that are close together have a smaller difference in measured values than those farther apart. The degree of spatial correlation is quantified with the experimental semivariogram, which measures correlation between measurements as a function of distance between the measurement points. Kriging computes an estimate best representing the spatial distribution of the measured values based on a semivariogram model that is fitted to the experimental semivariogram and a minimization of the estimation variance (or estimation error) at measured locations. Each estimate is accompanied by a corresponding standard error.

One of the key assumptions of kriging is that of stationarity (Isaaks and Srivastava, 1989, p. 349). Stationarity requires that the expected value (that is, the mean) of the data being estimated does not change when shifted in space, and that the modeled semivariogram is the same everywhere. This assumption is violated whenever there is a significant spatial trend in the measured values. For example, measured water-level elevations in the ESRP aquifer show a consistent, decreasing trend in the southwest direction (Lindholm and others, 1988). In such cases, nonstationarity can be accounted for in the data by use of a trend model. The trend model used for this report is a simple polynomial function (planar) fit to the data using linear regression[1]. Residuals are obtained by subtracting the trend from the measured data. Because the residuals should be stationary, kriging is applied to the residuals and the resulting estimate is added back to the trend to compute an estimate of the measured values. This method of kriging with a trend model is known in geostatistics as universal kriging (UK) (Pebesma, 2004).

---

[1] An external drift function using topographic elevation from a digital elevation model was also considered for this analysis. Kriging with an external drift (KED) assumes that the water table is a subdued replica of the topography. The relatively large permeability and low areal recharge rate of the ESRP aquifer indicate that the water table is essentially unrelated to the topography (Haitjema and Mitchell-Bruker, 2005, p. 786); thus, invalidating a key assumption of KED.

Kriging Formulation

A UK model was used for point estimates of water-level elevations in the ESRP aquifer. A general description of the kriging formulation is provided in many geostatistical texts such as Isaaks and Srivastava (1989, p. 278–322) and Kitanidis (1997, p. 125–127). A UK model is formulated here to describe water-level estimates in this study.

The UK model represents the water-level elevations as wavering about a deterministic function (or trend), and information about the scale and intensity of fluctuations about this trend is provided by a random function with zero mean and a correlation structure (Kitanidis, 1997, p. 120). In mathematical terms, this is expressed as:

$$z(s) = m(s) + e(s) \qquad (1)$$

where

s	is a pair of Cartesian coordinates describing the geographic location (point);
$z(s)$	is the median water-level elevation for the 2008 measurements at point s, in meters above NAVD 88;
$m(s)$	is the deterministic part of z at point s, in meters above NAVD 88; and
$e(s)$	is the stochastic part of z at point s, in meters.

A boldface algebraic symbol (such as, s in equation 1) is used to denote a vector quantity. The deterministic function $m(s)$, is called the trend and is defined as the expected value (E) of the water-level elevations, denoted by:

$$E[z(s)] = m(s). \qquad (2)$$

Trend is modeled as a linear polynomial function and is defined as the least-squares fit of a planar surface to the measured data, described by:

$$m(s) = \beta_0 + \beta_1 x(s) + \beta_2 y(s) \qquad (3)$$

where

$x(s)$	is the x coordinate (or easterly value) at point s, in meters;
$y(s)$	is the y coordinate (or northerly value) at point s, in meters;
β_0	is a deterministic unknown trend coefficient, in meters above NAVD 88; and
β_1, β_2	are deterministic unknown trend coefficients.

To obtain an estimate of z at a point s_0 (an estimation point) from water-level elevation measurements $z(s_1)$, $z(s_2)$,..., $z(s_n)$, the following are required:

1. The estimate is a linear function of measured data, that is:

$$\hat{z}(s_0) = \sum_{i=1}^{n} \lambda_i z(s_i) \tag{4}$$

where

$\hat{z}(s_0)$ is the estimate of z at point s_0,
$z(s_i)$ is the measurement of z at point s_i,
λ_i is the coefficient corresponding to well site i, and
n is the total number of measurement sites.

2. The estimate at measured points is unbiased, that is:

$$E\left[\hat{z}(s_0) - z(s_0)\right] = 0. \tag{5}$$

3. The estimated variance in square meters at point s_0 ($\sigma_{UK}^2(s_0)$) should be as small as possible, where variance is defined as:

$$\sigma_{UK}^2(s_0) = E\left[\left(\hat{z}(s_0) - z(s_0)\right)^2\right]. \tag{6}$$

The unbiased condition (equation 5), combined with the estimate in equation (4) and the trend model in equation (3) becomes:

$$E\left[\sum_{i=1}^{n} \lambda_i z(s_i) - z(s_0)\right] = 0 \tag{7}$$

$$\sum_{i=1}^{n} \lambda_i E\left[z(s_i)\right] - E\left[z(s_0)\right] = 0$$

$$\sum_{i=1}^{n} \lambda_i m(s_i) - m(s_0) = 0$$

$$\sum_{i=1}^{n} \lambda_i \left[\beta_0 + \beta_1 x(s_i) + \beta_2 y(s_i)\right] - \left[\beta_0 + \beta_1 x(s_0) + \beta_2 y(s_0)\right] = 0$$

$$\beta_0 \sum_{i=1}^{n} \lambda_i - \beta_0 + \beta_1 \sum_{i=1}^{n} \lambda_i x(s_i) - \beta_1 x(s_0) + \beta_2 \sum_{i=1}^{n} \lambda_i y(s_i) - \beta_2 y(s_0) = 0$$

$$\beta_0 \left(\sum_{i=1}^{n} \lambda_i - 1\right) + \beta_1 \left(\sum_{i=1}^{n} \lambda_i x(s_i) - x(s_0)\right) + \beta_2 \left(\sum_{i=1}^{n} \lambda_i y(s_i) - y(s_0)\right) = 0.$$

For this condition (equation 7) to hold for any values of β_0, β_1, and β_2 requires that:

$$\sum_{i=1}^{n} \lambda_i = 1, \ \sum_{i=1}^{n} \lambda_i x(s_i) = x(s_0), \ \text{and} \ \sum_{i=1}^{n} \lambda_i y(s_i) = y(s_0). \tag{8}$$

The estimated variance (equation 6), combined with the estimate in equation (4), becomes:

$$\sigma_{UK}^2\left(s_0\right) = E\left[\left(\sum_{i=1}^{n} \lambda_i\, z\left(s_i\right) - z\left(s_0\right)\right)^2\right]. \tag{9}$$

Adding and subtracting the trend (m) from the estimated variance (equation 9) gives

$$\sigma_{UK}^2\left(s_0\right) = E\left[\left(\sum_{i=1}^{n} \lambda_i\, z\left(s_i\right) - z\left(s_0\right) + m\left(s_0\right) - m\left(s_0\right)\right)^2\right] \tag{10}$$

$$= E\left[\left(\sum_{i=1}^{n} \lambda_i\, z\left(s_i\right) - \sum_{i=1}^{n} \lambda_i\, m\left(s_i\right) - z\left(s_0\right) + m\left(s_0\right)\right)^2\right]$$

$$= E\left[\left(\sum_{i=1}^{n} \lambda_i\left[z\left(s_i\right) - m\left(s_i\right)\right] - \left[z\left(s_0\right) - m\left(s_0\right)\right]\right)^2\right]$$

$$= E\left[\left(\sum_{i=1}^{n} \lambda_i\, e\left(s_i\right) - e\left(s_0\right)\right)^2\right]$$

$$= E\left[\left(\sum_{i=1}^{n} \lambda_i\, e\left(s_i\right) - e\left(s_0\right)\right)\left(\sum_{i=1}^{n} \lambda_i\, e\left(s_i\right) - e\left(s_0\right)\right)\right]$$

$$= E\left[\left(\sum_{i=1}^{n} \lambda_i\, e\left(s_i\right)\right)^2 - 2\sum_{i=1}^{n} \lambda_i\, e\left(s_i\right) e\left(s_0\right) + e\left(s_0\right)^2\right]$$

$$= E\left[\sum_{i=1}^{n}\sum_{j=1}^{n} \lambda_i \lambda_j\, e\left(s_i\right) e\left(s_j\right) - 2\sum_{i=1}^{n} \lambda_i\, e\left(s_i\right) e\left(s_0\right) + e\left(s_0\right)^2\right]$$

$$= \sum_{i=1}^{n}\sum_{j=1}^{n} \lambda_i \lambda_j\, E\left[e\left(s_i\right) e\left(s_j\right)\right] - 2\sum_{i=1}^{n} \lambda_i\, E\left[e\left(s_i\right) e\left(s_0\right)\right] + E\left[e\left(s_0\right)^2\right].$$

For random variables X and Y, the covariance (cov) between variables is defined as:

$$\mathrm{cov}\left(X,Y\right) = E\left[XY\right] - E\left[X\right]E\left[Y\right] = E\left[\left(X - E\left[X\right]\right)\left(Y - E\left[Y\right]\right)\right]. \tag{11}$$

Therefore, between any two points, s and s', the covariance of the residual component in square meters is denoted as $C_e(s, s')$. Rewriting equation (10) in terms of C_e gives:

$$\sigma_{UK}^2\left(s_0\right) = \sum_{i=1}^{n}\sum_{j=1}^{n} \lambda_i \lambda_j C_e\left(s_i,s_j\right) - 2\sum_{i=1}^{n} \lambda_i C_e\left(s_i,s_0\right) + C_e\left(s_0,s_0\right). \tag{12}$$

Where C_e is equal to the variance of e in square meters (σ_e^2) minus the semivariogram of e in square meters (γ_e), or:

$$C_e\left(s,s'\right) = C_e\left(s_0,s_0\right) - \gamma_e\left(s,s'\right) = \sigma_e^2 - \gamma_e\left(s,s'\right). \tag{13}$$

The estimated variance (equation 12) is then expressed using the semivariogram as:

$$\sigma_{UK}^2\left(s_0\right) = \sum_{i=1}^{n}\sum_{j=1}^{n} \lambda_i \lambda_j\left[\sigma_e^2 - \gamma_e\left(s_i,s_j\right)\right] - 2\sum_{i=1}^{n} \lambda_i\left[\sigma_e^2 - \gamma_e\left(s_i,s_0\right)\right] + \sigma_e^2. \tag{14}$$

Coefficients $\lambda_1, \lambda_2, \ldots, \lambda_n$ are estimated by minimizing the expression of equation (14) subject to the linear constraints of equation (8), or:

$$\underset{\lambda_1, \lambda_2, \ldots, \lambda_n \in R}{\text{minimize}} \sum_{i=1}^{n} \sum_{j=1}^{n} \lambda_i \lambda_j \left[\sigma_e^2 - \gamma_e \left(s_i, s_j \right) \right] - 2 \sum_{i=1}^{n} \lambda_i \left[\sigma_e^2 - \gamma_e \left(s_i, s_0 \right) \right] + \sigma_e^2 \tag{15}$$

subject to:

$$\sum_{i=1}^{n} \lambda_i = 1$$

$$\sum_{i=1}^{n} \lambda_i \, x \left(s_i \right) = x \left(s_0 \right)$$

$$\sum_{i=1}^{n} \lambda_i \, y \left(s_i \right) = y \left(s_0 \right)$$

where

R is the set of all real numbers.

Once the λ coefficients have been determined, they are substituted back into equations (4) and (14) to determine estimates of the water-level elevation and variance, respectively. The calculated standard error (σ_{UK}) in meters is equal to the square root of the estimated variance. Universal kriging requires that the semivariogram of the residual component (γ_e) be known beforehand.

Semivariogram Formulation

The semivariogram is used to characterize the degree of spatial correlation present in the data. The semivariogram is developed for e, the residual component of equation (1). The residual is obtained by combining equations (1) and (3):

$$e\left(s\right) = z\left(s\right) - m\left(s\right) = z\left(s\right) - \beta_0 - \beta_1 \, x\left(s\right) - \beta_2 \, y\left(s\right). \tag{16}$$

Coefficients β_0, β_1, and β_2 are estimated through linear regression analysis by minimizing the sum of the squared difference between the measured values (z) and the linear trend model (m), or:

$$\underset{\beta_0, \beta_1, \beta_2 \in R}{\text{minimize}} \sum_{i=1}^{n} \left[z\left(s_i\right) - \beta_0 - \beta_1 \, x\left(s_i\right) - \beta_2 \, y\left(s_i\right) \right]^2. \tag{17}$$

The mean is the expected value (E) of the residuals denoted by:

$$\mu_e\left(s\right) = E\left[e\left(s\right) \right]. \tag{18}$$

Kriging assumes that the residuals are mean-centered; therefore, the mean of the residuals is denoted by:

$$E\left[e(s)\right] = 0. \tag{19}$$

The semivariogram γ_e is defined as one-half the variance (var) of the difference between residuals at points s and s':

$$\gamma_e(s,s') = \frac{1}{2}\text{var}\left[e(s) - e(s')\right]. \tag{20}$$

For a random variable X with an expected value $E[X]$, the variance of X is defined as:

$$\text{var}(X) = E\left[\left(X - E[X]\right)^2\right] = E\left[X^2\right] - \left(E[X]\right)^2. \tag{21}$$

Therefore, equation (20) may be expressed as:

$$\begin{aligned}
\gamma_e(s,s') &= \frac{1}{2}\left\{ E\left[\left(e(s) - e(s')\right)^2\right] - \left(E[e(s) - e(s')]\right)^2\right\} \tag{22}\\
&= \frac{1}{2}\left\{ E\left[\left(e(s) - e(s')\right)^2\right] - \left(E[e(s)] - E[e(s')]\right)^2\right\}\\
&= \frac{1}{2}\left\{ E\left[\left(e(s) - e(s')\right)^2\right] - (0-0)^2\right\}\\
&= \frac{1}{2}E\left[\left(e(s) - e(s')\right)^2\right].
\end{aligned}$$

To facilitate the semivariogram estimation, it is assumed that the semivariogram depends only on the distance between pairs of measurement points, that is:

$$\gamma_e(s,s') = \gamma_e(h), \tag{23}$$
$$h = \sqrt{\left(x' - x\right)^2 + \left(y' - y\right)^2}$$

where

h	is the distance measured between point pairs s and s', in meters;
x, y	are coordinates where measurements were taken (s points); and
x', y'	are coordinates where measurements were taken (s' points).

The semivariogram (equation 22) may then be expressed as:

$$\gamma_e(s,s') = \frac{1}{2}E\left[\left(e(s) - e(s+h)\right)^2\right]. \tag{24}$$

The empirical semivariogram (or sample semivariogram) ($\hat{\gamma}_e$), a nonparametric estimate of the semivariogram, is computed by grouping $\gamma_e(h)$ values, which are in a given h interval (or bin). The squared difference in residuals is averaged for point pairs separated by a distance that is contained within the same bin (lag distance). Assuming isotropic conditions (the orientation of the linear segment that connects two points is neglected), the empirical semivariogram is defined as:

$$\hat{\gamma}_e(\tilde{h}) = \frac{1}{2N_h(\tilde{h})} \sum_{i=1}^{N_h} \left[e(s_i) - e(s_i + h) \right]^2, \quad \text{for } \tilde{h} - \frac{1}{2}\Delta h \leq h < \tilde{h} + \frac{1}{2}\Delta h, \tag{25}$$

$$\text{let } \tilde{h} = \left\{ \frac{1}{2}\Delta h, \frac{3}{2}\Delta h, \frac{5}{2}\Delta h, \frac{7}{2}\Delta h, \dots \right\}$$

where

$\tilde{h}$ is the lag distance (or the bin midpoint) at which the empirical semivariogram is computed, in meters;

$\hat{\gamma}_e(\tilde{h})$ is the empirical semivariogram, in square meters;

Δh is a constant bin width, in meters; and

N_h is the number of data pairs in each bin.

The empirical semivariogram is modeled with a continuous function that represents a theoretical semivariogram ($\hat{\gamma}_e$). A spherical model was selected for this report to represent the theoretical semivariogram, expressed as:

$$\hat{\gamma}_e(h) = \begin{cases} c, & \text{for } h > r \\ g + (c - g)\left[\frac{3}{2}\frac{h}{r} - \frac{1}{2}\left(\frac{h}{r}\right)^3 \right], & \text{for } 0 < h \leq r \\ 0, & \text{for } h = 0 \end{cases} \tag{26}$$

where

h is the lag distance in meters;

c is the sill (or upper bound of the semivariogram), in square meters;

g is the nugget (or the semivariance of the residual at a lag distance of zero), in square meters; and

r is the range (or lag distance at which the semivariogram reaches the sill), in meters.

The range indicates the distance over which data are correlated (either positively or negatively). Non-linear regression is used to fit the sill and range coefficients in the theoretical semivariogram (equation 26); whereas, the nugget is set by visual inspection. The regression is stated in the following optimization formulation:

$$\underset{c, r \in R}{\text{minimize}} \sum_{j=1}^{n_v} N_{h,j} \left[\hat{\gamma}_e\left(\tilde{h}_j\right) - \hat{\gamma}_e\left(\tilde{h}_j\right) \right]^2 \tag{27}$$

where

n_v is the number of points in the empirical semivariogram.

Semivariogram Development

The semivariogram (γ_e in equation 14) was developed using residual water-level elevations with respect to a regional spatial trend (equation 16); in this study, trend is represented as a planar function of coordinate variables. Regression analysis was used to estimate values for the coefficients of the planar trend model from the measured water-level elevations ($\beta_0 = 951.2$ m, $\beta_1 = 0.00142$, $\beta_2 = 0.00046$) (equation 17). An adjusted coefficient of determination [$\bar{R}^2$] equal to 0.86 (probability of an observed result arising by chance [p-value] = 2.2×10^{-16}) indicates that the trend model fits the data well and that the assumption of stationarity is valid. Wells in both monitoring networks ($n = 335$) were included in the analysis. Residuals calculated as the difference between the actual measurements and the trend range from -215.0 to 236.0 m, with a mean and standard deviation of 0.0 m and 39.1 m, respectively.

The empirical semivariogram of the residuals based on a constant bin width ($\Delta h = 10,000$ m) is shown in figure 5. The spatial separation distance to which point pairs are included in the semivariance estimates is 150 km (about half of the maximum separation distance between point pairs). There appears to be no nugget, indicating that continuity of the water-level elevation is high over short distances. Points in the empirical semivariogram are fit with a spherical model (theoretical semivariogram) with nugget fixed at zero ($g = 0$ m^2, $c = 1,949$ m^2, $r = 153,991$ m; coefficient of determination [R^2] = 0.82) (equation 26, fig. 5). Semivariance values less than the computed sample variance of the residual components ($\sigma_e^2 = 1,531$ m^2) are positively correlated and values greater than the sample variance are negatively correlated.

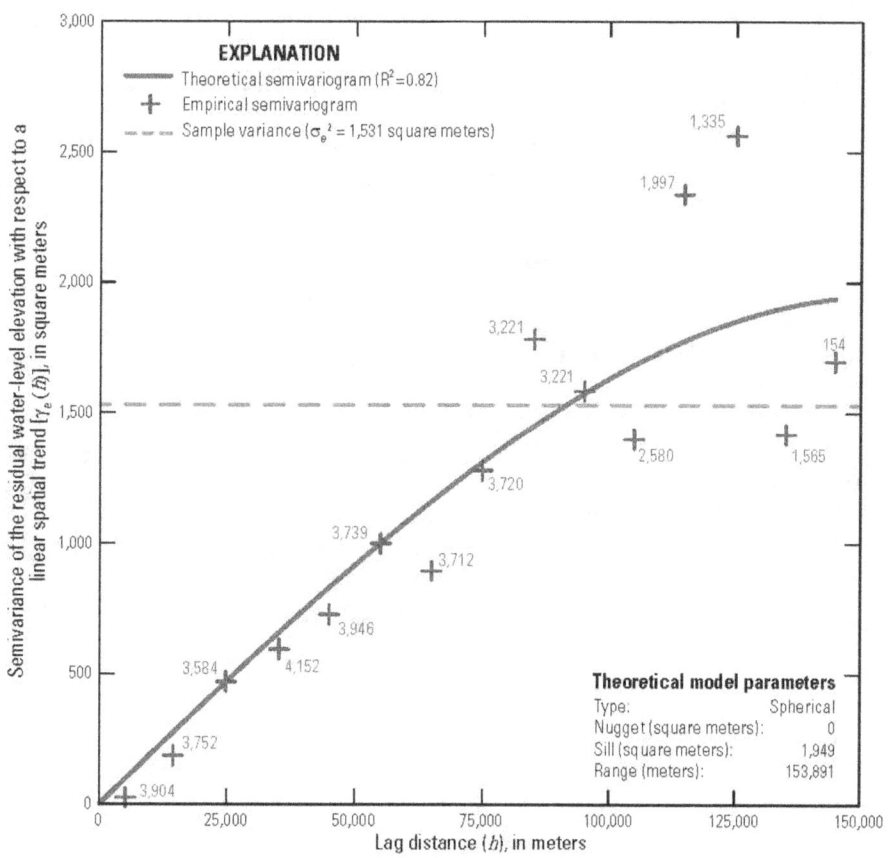

Figure 5. Semivariogram analysis of water-level elevation residuals after subtraction from trend. Numbers next to symbols refer to the number of sampled data pairs in a lag-distance interval (N_h).

Kriging Implementation

Kriging analysis was used to interpolate water levels at points in a uniform grid (kriging grid) oriented north-south and east-west with square blocks. The spatial resolution of a kriging grid is specified using the length of a square block side in the grid (ℓ). A spatial resolution of ℓ equal to 500 m is used for all map figures in this report. Selection of this grid resolution was based on the inherent spatial variability of observation wells in the ESRP. Three simulations of the water-table surface were run using water-level elevations measured from wells in: (1) the Co-op and USGS-INL networks ($n = 335$), (2) the Co-op network ($n = 166$), and (3) the USGS-INL network ($n = 171$). The area of analysis for simulations (1) and (2) is defined by the generalized boundary of the ESRP (fig. 1) (number of nodes in the kriging grid [n_n] = 112,325; area = 28,081 km^2). The estimated water-table surface (at 50-m contour intervals) based on water-level measurements from wells in both networks is shown in figure 6. The interpolated water table (ranging from 872.9 to 1,742.8 m) is consistent with other water-table maps constructed for the ESRP aquifer for 1928–30, 1956–58, and 1980 (Stearns and others, 1938; Mundorff and others, 1964; Lindholm and others, 1988). The estimated water-table surface (at 50-m contour intervals), based on water-level measurements from wells only in the Co-op network, is shown in figure 7A. A comparison of this water-table map with the map derived from wells in both networks (fig. 6) shows large differences in the shape of the water-table contours beneath the INL and vicinity.

For kriging of water-level elevations measured from wells in the USGS-INL network (simulation 3), the area of analysis is defined by the part of the ESRP beneath the INL and vicinity ($n_n = 18,394$; area = 4,599 km^2). The interpolated water-table surface is described using a 5-m contour interval (fig. 8A); the kriging-based water-table map is consistent with a March–May 2008 water-table map for this area constructed previously using a multilevel B-splines interpolation technique (Fisher and Twining, 2011, fig. 4).

Prediction Uncertainty

An advantage of kriging (over other interpolation algorithms such as IDW and splines) is that every estimate of the water-level elevation is accompanied by a corresponding measure of the uncertainty associated with the estimate (that is, the standard error, or square root of the estimated variance; equation 14). Values of standard error are basically a scaled version of the distance to the nearest measurement point; for example, standard error is zero at measured points and increases as the density of the monitoring network decreases. Standard error for kriging based on water levels measured from wells in the Co-op network ranged from 1.5 to 31.4 m (fig. 7B). The west-central part of the ESRP shows significant uncertainty resulting from a scarcity of Co-op network wells in this area. Standard error for kriging based on water levels measured from wells in the USGS-INL network ranged from 0.6 to 31.0 m (fig. 8B); uncertainty is greatest in the southeast and northeast corners of the kriging grid, areas without USGS-INL network wells where extrapolated predictions are subject to greater uncertainty.

Cross-Validation

Cross-validation is a specialized resampling procedure, used here to (1) indicate if there are significant flaws in the kriging model, and (2) identify locations where the water table is most dynamic. The resampling procedure (David, 1976; Delfiner, 1976), known as leave-one-out cross-validation, uses all water-level measurements to estimate the spatial trend (equation 3) and the theoretical semivariogram model (equation 26). Leave-one-out cross-validation removes one site from the data set (well sites in both monitoring networks are included in this data set, $n = 335$) and estimates the water-level elevation at that location by kriging with the remaining data. The estimation error, that is, the difference between the actual (z) and estimated values (z^*) at the location of the omitted site ($z - z^*$), is then computed (table 5). This procedure is repeated for each site in the data set.

Water table based on the Co-op and USGS-INL networks

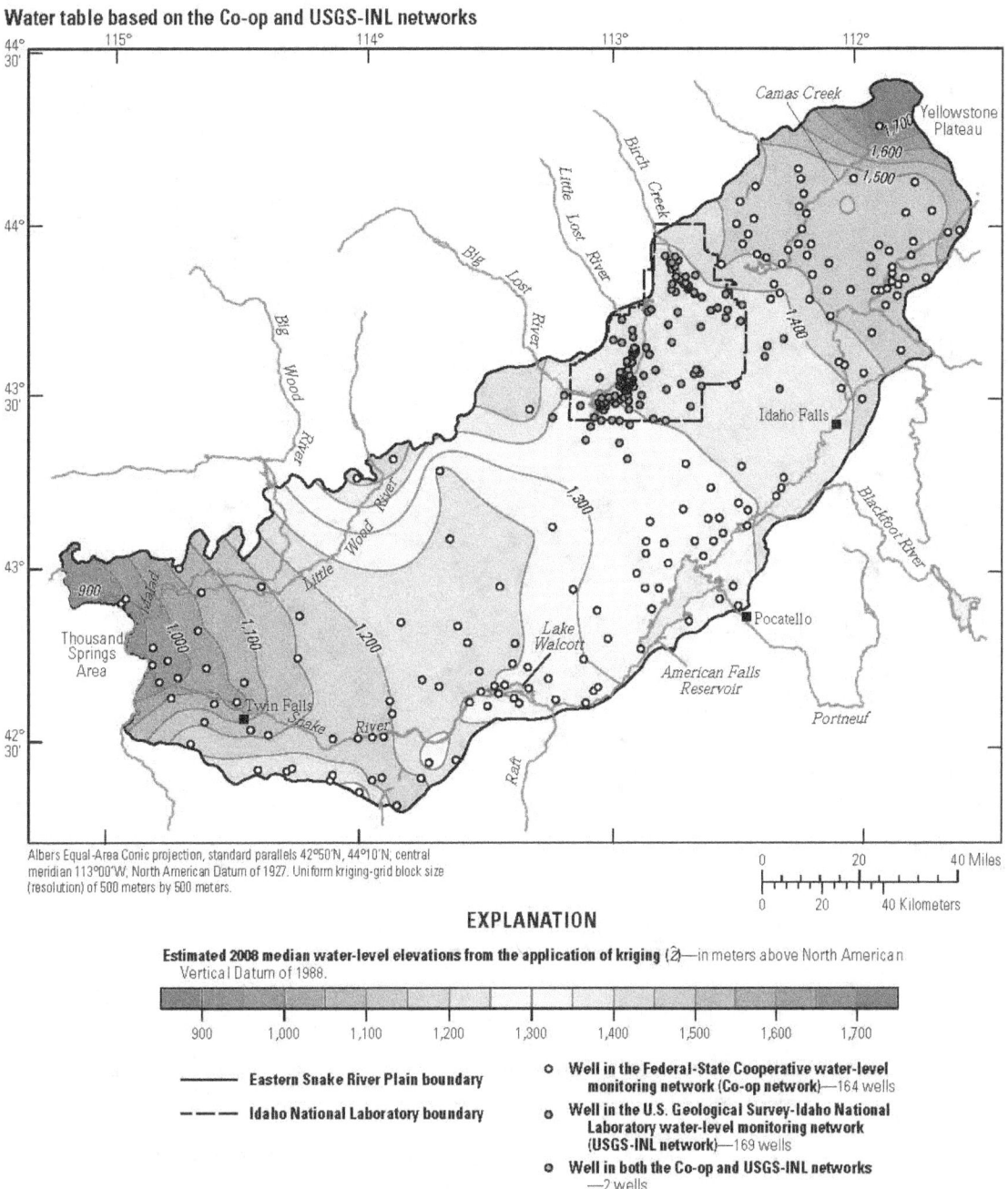

Albers Equal-Area Conic projection, standard parallels 42°50'N, 44°10'N, central meridian 113°00'W, North American Datum of 1927. Uniform kriging-grid block size (resolution) of 500 meters by 500 meters.

0   20   40 Miles

0   20   40 Kilometers

EXPLANATION

Estimated 2008 median water-level elevations from the application of kriging (ẑ)—in meters above North American Vertical Datum of 1988.

900   1,000   1,100   1,200   1,300   1,400   1,500   1,600   1,700

——— Eastern Snake River Plain boundary

– – – Idaho National Laboratory boundary

○ Well in the Federal-State Cooperative water-level monitoring network (Co-op network)—164 wells

◉ Well in the U.S. Geological Survey-Idaho National Laboratory water-level monitoring network (USGS-INL network)—169 wells

◉ Well in both the Co-op and USGS-INL networks —2 wells

Figure 6. Estimated water table from kriging of water levels measured in the Federal-State Cooperative water-level monitoring network and U.S. Geological Survey-Idaho National Laboratory water-level monitoring network, eastern Snake River Plain, Idaho.

***A.* Water table based on the Co-op network**

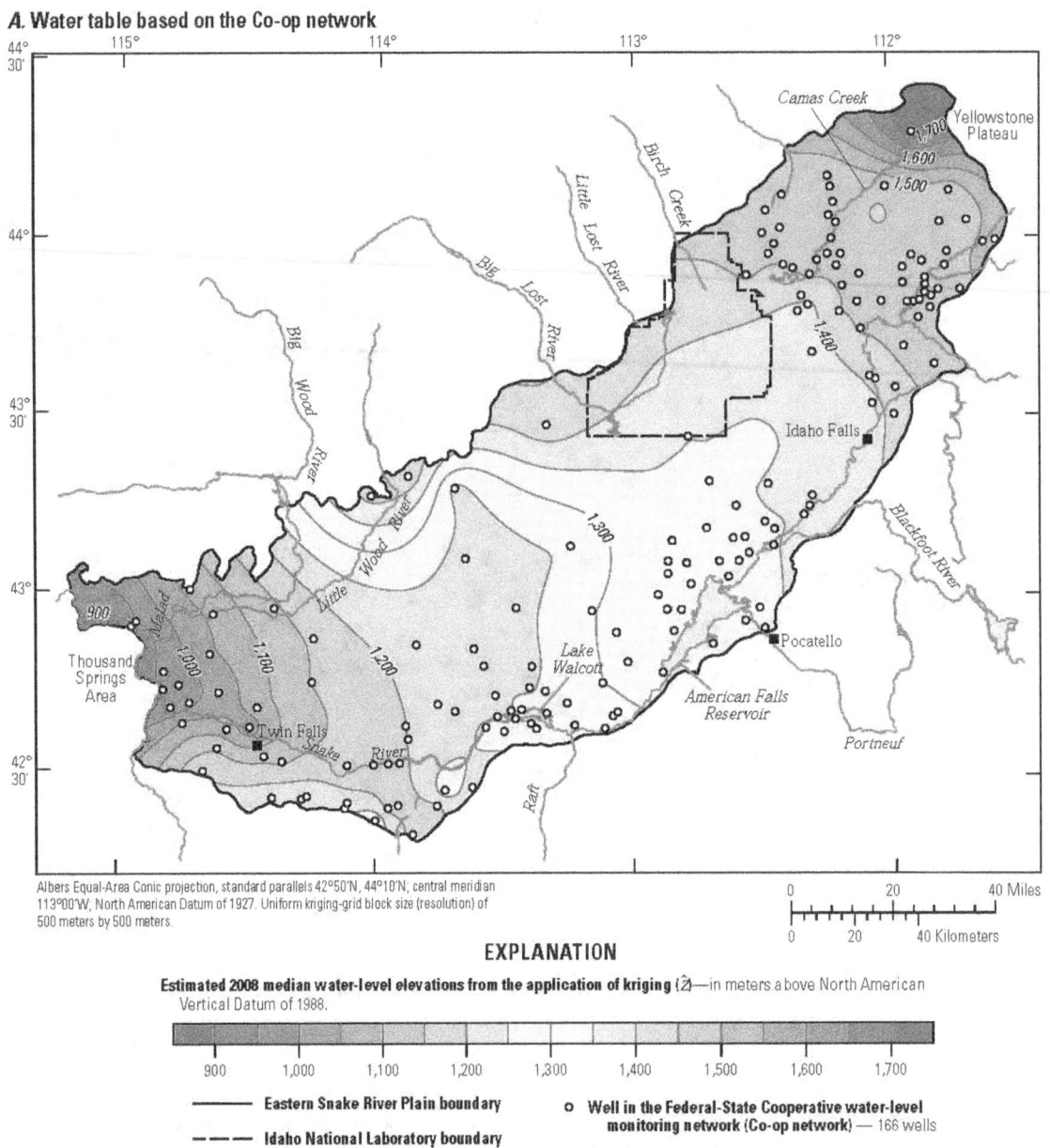

Albers Equal-Area Conic projection, standard parallels 42°50'N, 44°10'N; central meridian
113°00'W; North American Datum of 1927. Uniform kriging-grid block size (resolution) of
500 meters by 500 meters.

EXPLANATION

Estimated 2008 median water-level elevations from the application of kriging (*A***)**—in meters above North American
Vertical Datum of 1988.

| 900 | 1,000 | 1,100 | 1,200 | 1,300 | 1,400 | 1,500 | 1,600 | 1,700 |

——— Eastern Snake River Plain boundary

– – – Idaho National Laboratory boundary

○   Well in the Federal-State Cooperative water-level
    monitoring network (Co-op network) — 166 wells

Figure 7.   (*A*) Estimated water table and (*B*) uncertainty from kriging of water levels measured in the Federal-
State Cooperative water-level monitoring network, eastern Snake River Plain, Idaho.

B. Uncertainty based on the Co-op network

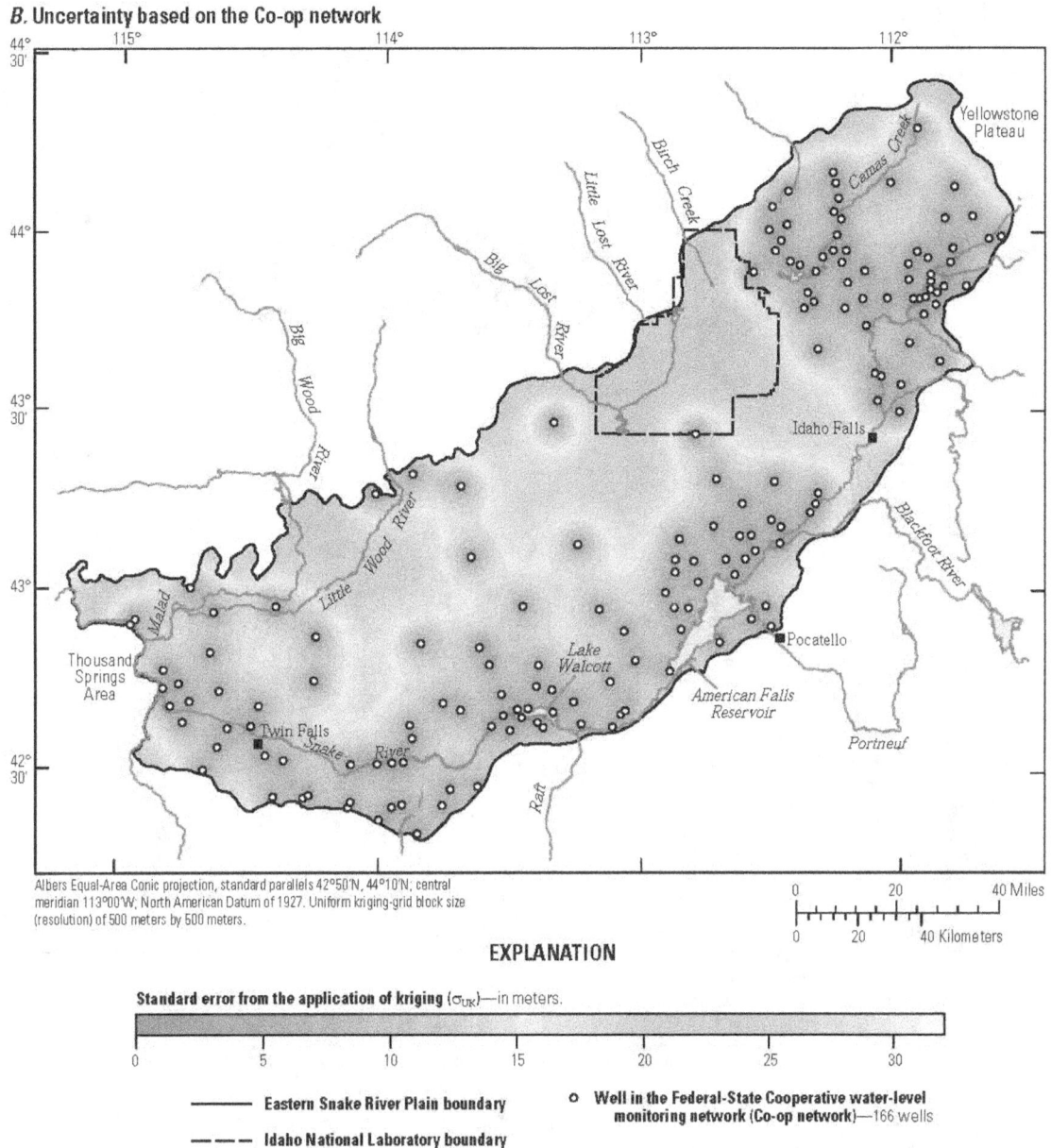

Albers Equal-Area Conic projection, standard parallels 42°50'N, 44°10'N; central meridian 113°00'W; North American Datum of 1927. Uniform kriging-grid block size (resolution) of 500 meters by 500 meters.

EXPLANATION

Standard error from the application of kriging (σ_{UK})—in meters.

| 0 | 5 | 10 | 15 | 20 | 25 | 30 |

——— Eastern Snake River Plain boundary

– – – Idaho National Laboratory boundary

o Well in the Federal-State Cooperative water-level monitoring network (Co-op network)—166 wells

Figure 7.—Continued

A. Water table based on the USGS-INL network

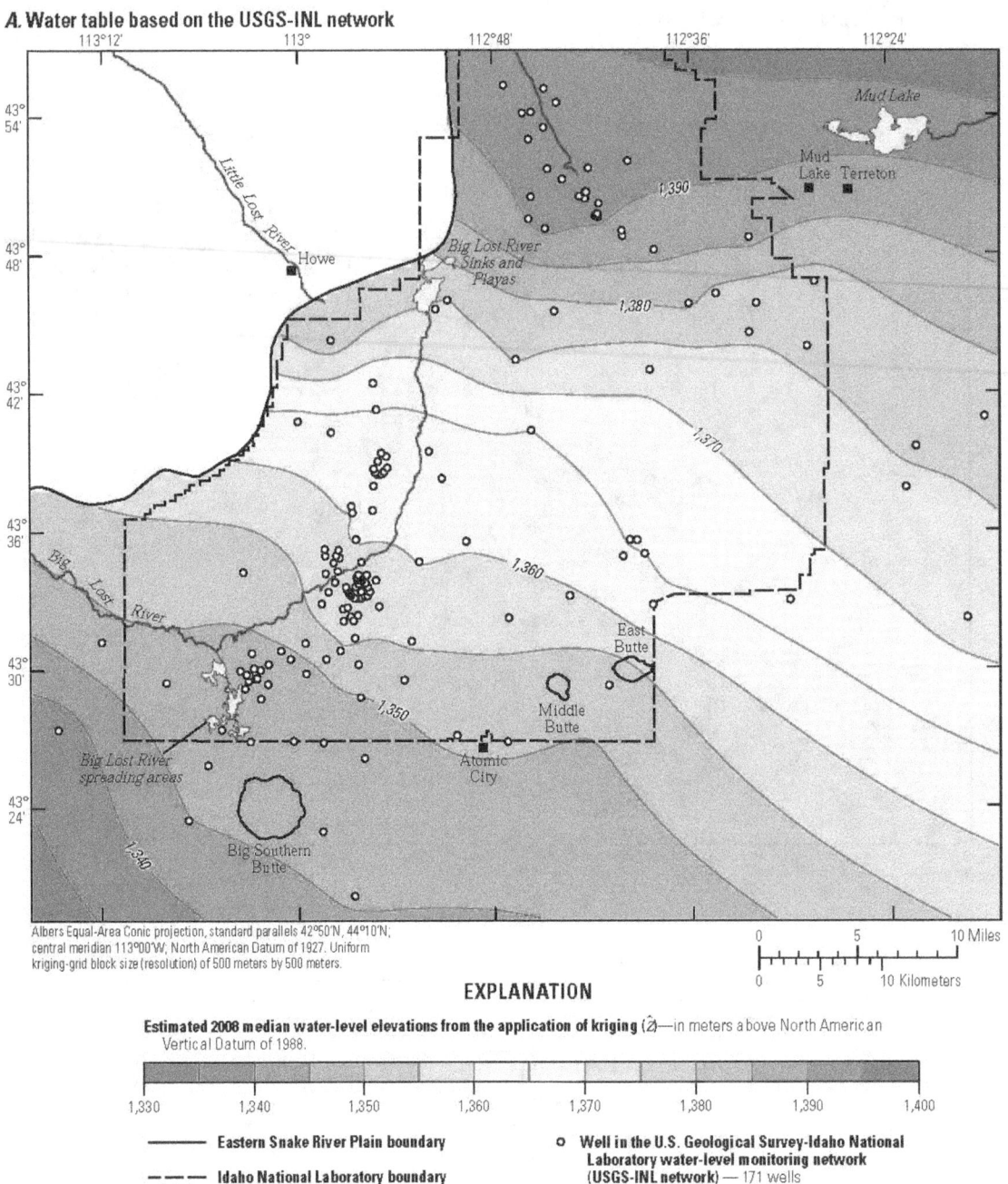

Figure 8. (*A*) Estimated water table and (*B*) uncertainty from kriging of water levels measured in the U.S. Geological Survey-Idaho National Laboratory water-level monitoring network, Idaho National Laboratory and vicinity, Idaho.

B. Uncertainty based on the USGS-INL network

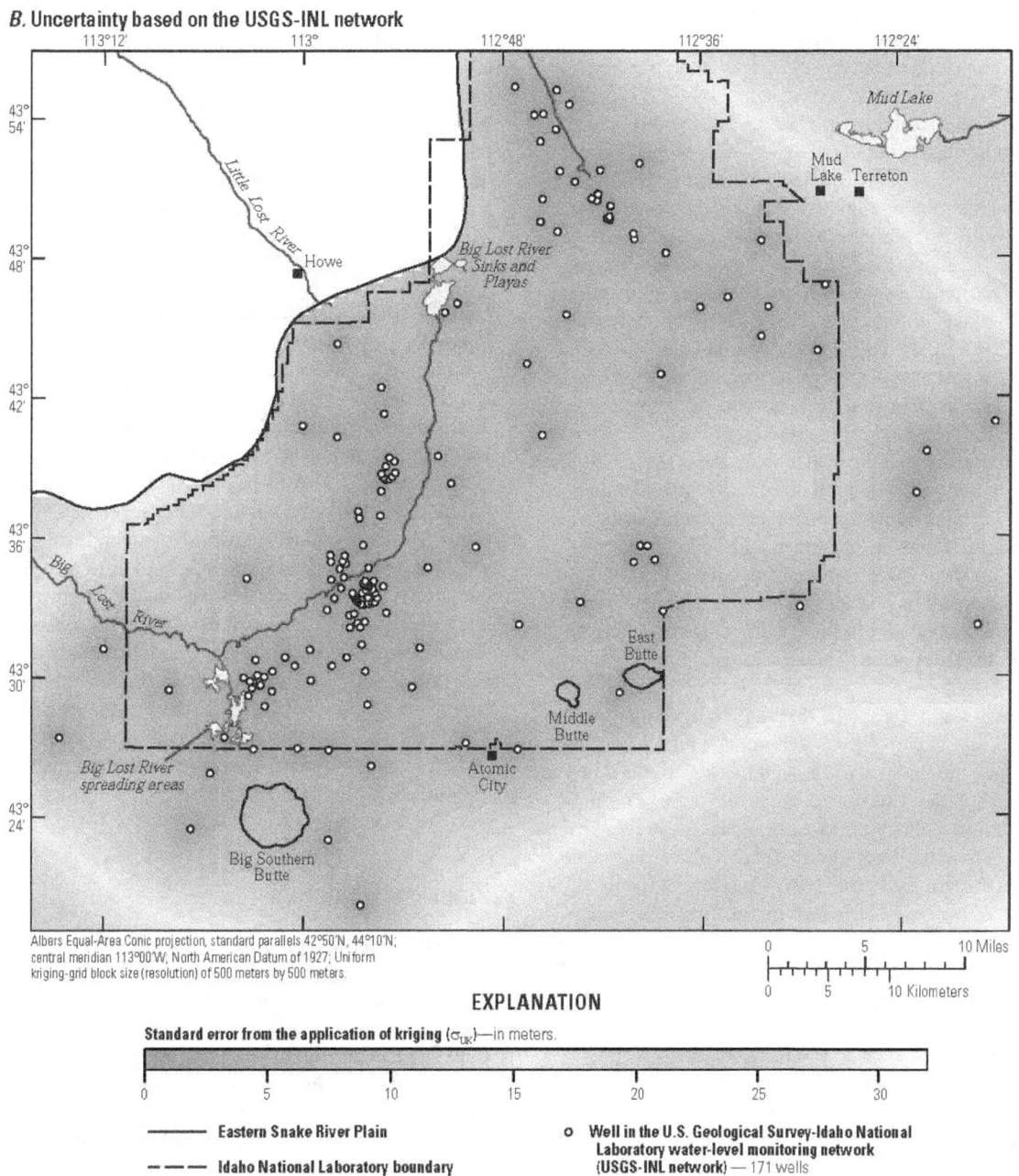

Albers Equal-Area Conic projection, standard parallels 42°50'N, 44°10'N;
central meridian 113°00'W; North American Datum of 1927; Uniform
kriging-grid block size (resolution) of 500 meters by 500 meters.

0   5   10 Miles

0   5   10 Kilometers

EXPLANATION

Standard error from the application of kriging (σ_{UK})—in meters.

0   5   10   15   20   25   30

——— Eastern Snake River Plain

– – – Idaho National Laboratory boundary

o   Well in the U.S. Geological Survey-Idaho National
     Laboratory water-level monitoring network
     (USGS-INL network) — 171 wells

Figure 8.—Continued

A scatter plot of measured values and estimated values in the leave-one-out cross-validation analysis is shown in figure 9A. The slope of the regressed line (black dashed line, slope = 1.1 meters per meter [m/m], R^2 = 0.97) is about 45 degrees (gray solid line), indicating that the estimates are not conditionally biased (Philip and Kitanidis, 1989, p. 862). Conditional bias occurs when the kriging estimates do not have the same variability (standard deviation) as the measured values.

Figure 9B shows a scatter plot of the estimation error versus the estimated values of water-level elevation at the observation wells. Values of estimation error range from -144.7 to 267.5 m, with a mean and standard deviation of 0.7 m and 23.6 m, respectively. The mean estimation error is small (ideally zero), indicating an absence of systematic errors that could lead to biased estimations from the kriging model. Estimation error is equally scattered around a horizontal line and independent of the magnitude of the estimated values (R^2 = -0.003). This indicates stationarity may be assumed for the residual values (e). Points outside the 95-percent confidence interval (plus or minus twice the standard deviation) indicate that either the measurement is incorrect or the area where the observation well is located requires a denser network of wells (Theodossiou and Latinopoulos, 2006, p. 997).

An analysis of the spatial distribution of estimation error identifies (1) important observations for constructing the water-table map (such as isolated measurements that are different and distant from surrounding observations), and (2) confounding data that is in areas of high data density with water levels much different from other local measurements. Estimation error is shown in figure 10. Many of the locations of larger positive estimation error (green circles) coincide with areas of rapid change in the water table elevation. For example, the largest green circles at observation wells 2, 98, 103, 105, and 166 are in areas of steep hydraulic gradients along the margin of the ESRP. These wells are all paired with a band of nearby wells with relatively large negative estimation errors (red circles) that define the locations where the hydraulic gradients rapidly flatten. This type of pairing is also observed near the 1,440 m water-table contour where a band of relatively large green circles is just upgradient of a band of relatively large red circles (fig. 10). This area of steep hydraulic gradient coincided with changes in aquifer transmissivity near Mud Lake (fig. 2) (Lindholm and others, 1988). Relatively large estimation errors (both positive and negative) also can occur at observation wells in sparsely populated areas of the monitoring network. For example, the large estimation error at well 87 (-44.93 m) probably is because of its relative isolation within the network (about 21 km from the nearest neighboring well) and may indicate a need for higher network density in this area.

In areas where the estimation errors are relatively large and apparently random (that is, not because of the banding or isolation), the data are possibly erroneous. Alternatively, these relatively large estimation errors can be explained by extreme local abnormalities in the water table resulting from irrigation pumping, deep-percolation return flow beneath irrigated fields, or both. For example, estimation errors are relatively large and random in irrigated areas in the northeast and southwest parts of the ESRP (figs. 1 and 10).

Optimization of Water-Level Monitoring Networks

Each water-level monitoring network in the ESRP is optimized separately. Individually optimizing each network enables managers to adapt the results of this study into future network designs. However, water-table maps typically are interpolated from all available water-level data, with little-to-no distinction made as to which network the observation well belongs (other than to select wells from networks that maintain a consistent level of data quality). A disadvantage of individually optimizing each network is the omission of data from other networks; the exclusion of this data can result in data redundancy where network coverage overlaps and when data from multiple networks are used in the interpolation. To avoid this issue, network managers are encouraged to combine resources and to optimize a single monitoring network. Optimization of combined networks in the ESRP is beyond the scope of this report.

Planning Objective

The planning objective for the water-level monitoring networks is to reduce total monitoring costs by removing wells from the original network because they add little or no information characterizing the water table. In this study, equal monitoring costs are assumed for each well. Although the validity of this assumption is untrue (for example, travel time can account for large variability in monitoring costs), it permits wells to be evaluated exclusively through a geostatistical analysis of the water-level elevation measurements. An estimate of the true cost savings for an optimized monitoring network is beyond the scope of this study; however, decreases in the total number of wells in a monitoring network typically will result in a reduction of total monitoring costs.

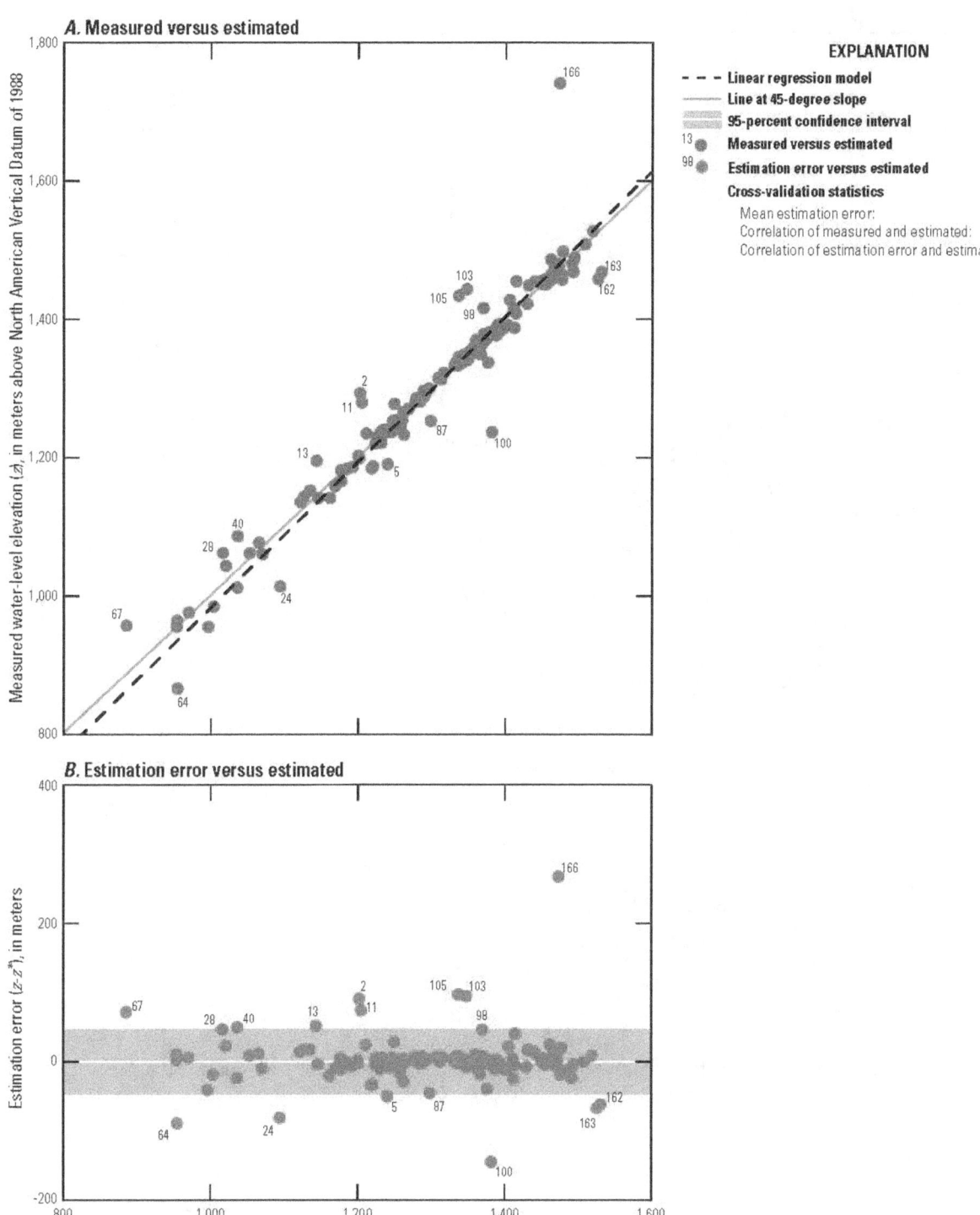

Figure 9. Scatter plots showing (*A*) measured and estimated water-level elevations, and (*B*) estimation error and estimated water-level elevations, from leave-one-out cross-validation, eastern Snake River Plain, Idaho. Map numbers included for sites located outside the 95-percent confidence interval of estimation error.

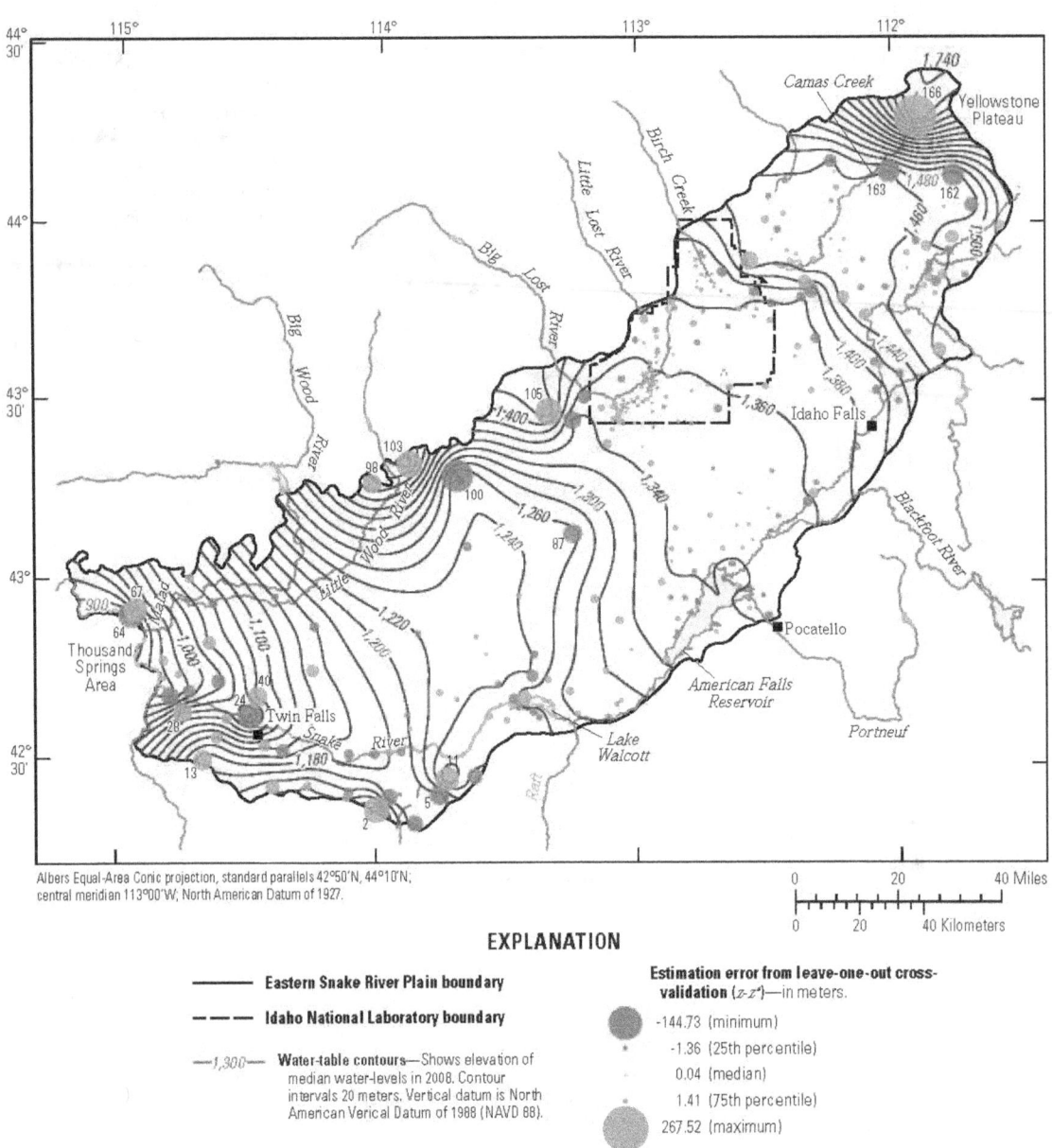

Albers Equal-Area Conic projection, standard parallels 42°50'N, 44°10'N; central meridian 113°00'W; North American Datum of 1927.

0   20   40 Miles

0   20   40 Kilometers

EXPLANATION

———— Eastern Snake River Plain boundary

– – – – Idaho National Laboratory boundary

—*1,300*— Water-table contours—Shows elevation of median water-levels in 2008. Contour intervals 20 meters. Vertical datum is North American Verical Datum of 1988 (NAVD 88).

Estimation error from leave-one-out cross-validation (z-z^*)—in meters.

-144.73 (minimum)

-1.36 (25th percentile)

0.04 (median)

1.41 (75th percentile)

267.52 (maximum)

Figure 10. Estimation errors from leave-one-out cross-validation, eastern Snake River Plain, Idaho. Map numbers included for sites located outside the 95-percent confidence interval of estimation error.

Design Criteria

To accomplish the established objective of the monitoring networks, water levels should be measured at well sites selected to satisfy the following design criteria:

1. The total number of wells in the optimized monitoring network is fixed and based on a user-defined number of wells to remove from the original network (n_r). Selecting an appropriate value for n_r is a management decision and typically requires a cost-benefit analysis. To assist decision makers, optimal monitoring networks corresponding to a variety of n_r values are included in this report.

2. Standard error from the application of kriging should be as small as possible. For example, removal of an observation well from an area of the monitoring network where few wells exist typically would result in a large increase in the interpolation error (defined as the mean standard error for all points [or nodes] in the kriging grid); therefore, this well would not likely be removed.

3. The difference between measured and estimated water-level elevations indicates the importance of an observation well for simulating the water-table surface. For example, well sites with smaller differences are less important because their exclusion from the existing monitoring network would have little-to-no effect on the distribution of water levels in the kriged surface. In comparison, water-level distribution is much more sensitive to observation wells in areas where differences between measured and estimated values are large. Localized water-table features are preserved by retaining wells where the difference between measured and estimated values is large.

4. The variability of water-level measurements is preserved across time. Observation wells with hydrographs showing prominent seasonal fluctuations and long-term trends are important for understanding the availability and sustainability of groundwater resources. For example, observation wells along the margin of the ESRP should show a stronger seasonality associated with recharge, whereas wells in the center of the plain should have a damped response. Near pumping areas, wells may have a different type of seasonality, with low water levels coinciding with peak demand. Informed decisions about the aquifer often require an understanding of seasonality. The standard deviation was used to identify the temporal variations in water-level measurements. Sites with small standard deviations are more likely to be excluded from a monitoring network than sites with large standard deviations.

5. The measurement error of water-level elevations should be as small as possible. Well sites with large measurement error are more likely to be excluded from the network. Measurement error is estimated as the sum of the accuracy to which the measurement point elevation is known plus the accuracy of the measurement method.

Each of these criteria (with the exception of design criterion (1), the number of sites to remove) was converted to a mathematical metric, and the metrics were combined into a single multi-objective function that was used to identify a water-level monitoring network satisfying the design criteria as much as possible.

Multi-Objective Problem Formulation

The multi-objective problem is formulated as a single-objective optimization where a weighted combination of the design criteria is minimized. In mathematical terms, this is expressed as:

$$\underset{x_1,x_2,\ldots,x_m \in \mathbf{Z}}{\text{minimize}}\, F'(x) = \begin{cases} F(x), & \text{for valid "decision variables"} \\ P(x), & \text{otherwise} \end{cases} \quad (28)$$

where

$\mathbf{Z}$	is the set of all integers;
x	is the decision variables;
F'	is the "fitness" function, in meters;
F	is the weighted-sum objective function, in meters; and
P	is the "penalty" function, in meters.

The minimum fitness value corresponds to the optimal monitoring network. The fitness function is dependent on the decision variables, a vector of integer values used to identify observation wells in the existing monitoring network that will not be included in the reduced network. The purpose of the optimization solver is to find values of x that minimize the fitness value.

For valid combinations of decision variables, the fitness value is calculated using the function F, given by:

$$F(x) = w_1 f_1(x) + w_2 f_2(x) + w_3 f_3(x) + w_4 f_4(x) \quad (29)$$

where

f	is the individual objective function, in meters; and
w	is the weighting coefficient.

All the design criteria except for the management decision of how many wells to remove from the existing monitoring network (n_r) are quantified by each of the individual objective functions: $f_1, f_2, f_3,$ and f_4. The relative influence of each criterion may be established by varying the associated weights: $w_1, w_2, w_3,$ and w_4.

For functions f_1 and f_2, kriging estimates are based on the reduced monitoring network, which includes all original network wells that are not identified by the optimization solver for removal. In mathematical terms, measurement points in the reduced network are described using set-builder notation as:

$$s_i \quad \text{for all } i \in \{i \in \mathbf{Z} \mid 1 \geq i \leq n_e \text{ and } i \notin \boldsymbol{x}\} \quad (30)$$

where

$\quad n_e \quad$ is the number of observation wells in the existing monitoring network.

The function f_1 is the metric selected to minimize the standard error, and is defined as:

$$f_1 = \frac{1}{n_n} \sum_{i=1}^{n_n} \sqrt{\sigma_{UK}^2 \left(s_{n,i} \right)} \quad (31)$$

where

$\quad s_{n,i} \quad$ is the spatial coordinates corresponding to node i in the kriging grid;

$\quad n_n \quad$ is the number of nodes in the kriging grid; and

$\quad \sigma_{UK}^2(s_{n,i}) \quad$ is the estimated variance at point $s_{n,i}$ based on the reduced monitoring network (equation 14), in square meters.

The summation in f_1 is over all nodes in the kriging grid (equation 31). Because standard error in each node depends on the proximity of nearby observation wells, removal of wells from regions that have sparser data increases standard error more than at nodes that are close to other supporting data.

The function f_2 is the metric selected to minimize the root-mean-squared-error (RMSE), and is defined as:

$$f_2 = \sqrt{\frac{1}{n_r} \sum_i \left[z(s_i) - \hat{z}(s_i) \right]^2} \quad (32)$$

$$\text{for all } i \in \{i \in \mathbf{Z} \mid 1 \geq i \leq n_e \text{ and } i \in \boldsymbol{x}\}$$

where

$\quad s_i \quad$ is the spatial coordinates at well site i;

$\quad z(s_i) \quad$ is the median water-level elevation at point s_i for the 2008 measurements, in meters; and

$\quad \hat{z}(s_i) \quad$ is the estimated water-level elevation at point s_i based on measurements in the reduced monitoring network, in meters (equation 4).

The summation in f_2 is over well sites selected for removal from the existing monitoring network (equation 32). Removal of well sites with small differences between measured and estimated values decreases the RMSE more than removing wells with large differences.

The function f_3 is the metric selected to preserve the variability of water-level measurements across time, and is defined as

$$f_3 = \frac{1}{n_r} \sum_i \sigma_z(s_i) \quad \text{for all } i \in \{i \in \mathbf{Z} \mid 1 \geq i \leq n_e \text{ and } i \in \boldsymbol{x}\} \quad (33)$$

where

$\quad \sigma_z(s_i) \quad$ is the standard deviation of all water-level elevation measurements collected at point s_i across the period of record (duration varies at each well site), in meters.

The summation in f_3 is over well sites selected for removal from the existing monitoring network (equation 33). Removing wells with small standard deviations preserves the variability more than removing wells with large standard deviations.

The function f_4 is the metric selected to minimize the mean measurement error, and is defined as:

$$f_4 = \frac{1}{n - n_r} \sum_i \varepsilon_z(s_i) \quad \text{for all } i \in \{i \in \mathbf{Z} \mid 1 \geq i \leq n_e \text{ and } i \notin \boldsymbol{x}\} \quad (34)$$

where

$\quad \varepsilon_z(s_i) \quad$ is the mean measurement error of z at point s_i estimated for the 2008 measurements, in meters.

The summation in f_4 is over well sites in the reduced monitoring network (equation 34). As expected, removing wells with large measurement errors will decrease the mean measurement error for the reduced monitoring network.

A penalty function, P, is used to penalize combinations of decision variables that are non-unique (that is, when $\boldsymbol{x}$ contains duplicate values) by setting the fitness value artificially large (equation 28). The penalty function is given by:

$$P(\boldsymbol{x}) = C \, n_{dup} \quad (35)$$

where

$\quad C \quad$ is the penalty coefficient, in meters; and

$\quad n_{dup} \quad$ is the number of duplicate wells in $\boldsymbol{x}$.

The linear dependency of P on n_{dup} indicates that, for networks with too few wells, the penalty value is proportional to the number of missing wells. For example, a network that is missing only 1 well would be penalized less than a network that is missing 20 wells. This helps to reduce the number of calls to the penalty function. The penalty coefficient, C, is selected to be arbitrarily large compared with the maximum possible F value. The high computational costs associated with kriging an infeasible network are circumvented by calling the low-cost penalty function.

Genetic Algorithm

A genetic algorithm (GA) (Holland, 1975) is used to find the best fitness value (that is, the minimum F' value in equation 28). GAs are adaptive heuristic search algorithms that mimic the mechanics of natural selection and survival of the fittest, and are well suited for solving combinatorial optimization problems in which there is a large set of candidate solutions. Koza (1992, p. 18) provides the following definition of a GA:

> "The *genetic algorithm* is a highly parallel mathematical algorithm that transforms a set (*population*) of individual mathematical objects (typically fixed-length character strings patterned after *chromosome* strings), each with an associated *fitness* value, into a new population (i.e., the next *generation*) using operations patterned after the Darwinian principle of reproduction and survival of the fittest and after naturally occurring genetic operations (notably sexual recombination)."

In GA terminology, the array of decision variables (or string of "genes") in the optimization problem is called a "chromosome", which for the current problem defines the set of observation wells being considered for removal from the existing monitoring network (x). The integer values used to identify wells in the original network (map numbers in table 5) are coded in the chromosome as fixed-length binary strings using Gray encoding (Gardner, 1986). A chromosome represents a unique solution in the solution space, the collection of all possible (or "candidate") solutions to the optimization problem. In this study, a chromosome describes a single design solution for the reduced water-level monitoring network (that is, well sites to exclude from the existing monitoring network). Each design solution (a particular set of wells uniquely defined in the chromosome) is assigned a fitness value (F'), which summarizes how well the particular set of wells meets the overall design objectives (described in the section, Design Criteria). The GA operates on a collection of chromosomes called a "population"; where the number of chromosomes in a population is expressed as n_{pop}.

An implementation of the GA begins with a population of random chromosomes (that is, suitable sets of randomly selected observation wells). During each "generation" (iteration) of the GA, the fitness of every chromosome in the population is computed. A subset of chromosomes is selected from the population based on their having superior fitness values, and is copied to a new population. This ensures that the best solutions can survive to the end of the GA run. The small part of the population that is guaranteed to survive to the next iteration is called "elitism" (r_e). The rest of the new population has many chromosomes modified from the present population.

Operators used to modify chromosomes include "crossover" (also called recombination) and "mutation". Crossover is the process of combining part of the data from two "parent" chromosomes to produce two new "child" chromosomes. Parents are selected from the present population using a linear-rank selection method. In linear-rank selection, chromosomes are ranked from best fitness value (rank = 1) to worst fitness value (rank = n_e), and are selected with a probability that is linearly proportional to their ranking. Part of the data from each parent is combined using single-point crossover— one crossover point on both parents' chromosome strings (decoded as integer values) is selected. All data from the beginning of the chromosome to the crossover point is swapped between the parents' chromosomes to produce the child chromosomes. There is a chance of introducing duplicate well sites into the children during crossover. To discourage this from happening, a child with duplicates is aborted, a new crossover point is randomly selected, and again data are swapped between parents. A limit is placed on the number of times a child chromosome can be aborted (n_a) to avoid biasing the location of the crossover point. If n_a is exceeded, the infeasible child chromosome containing duplicates survives and its associated fitness value is penalized (equation 35). The probability that crossover will occur between two parents is called the "crossover probability" (p_c). If there is no crossover, child chromosomes will be an exact copy of their parents. After crossover, mutation takes place.

Mutation is the process of altering a gene on the child chromosome. This mutation can result in the possible removal of a well site, thereby maintaining genetic diversity in the population from one iteration to the next. The location of the gene (or well site) on the child chromosome that will be mutated is randomly selected. The modified value for the mutated gene is randomly selected from the set of wells not already included in the child chromosome. The probability that a mutation will occur on a child chromosome is called the "mutation probability" (p_m). After crossover and mutation, the child chromosomes are copied to the new population.

New parents are selected for each new pair of children, and the process continues until the new population is filled. The new population is then used in the next iteration of the algorithm, with successive iterations producing smaller values of F'. For subsequent iterations, the fitness calculation is only needed for chromosomes derived from crossover, mutation operations, or both. The algorithm terminates based on the following criteria: (1) the maximum number of iterations (n_{iter}) is reached, or (2) the maximum number of consecutive iterations without any improvement in the best fitness value (n_{run}) is exceeded. Global optimality is not guaranteed with either of these stopping criteria; however, the probability of attaining the global optimum increases with an increase in the magnitude of n_{iter} or n_{run} values.

Computer Software

Computer software used to process data, to perform kriging analysis, to optimize the water-level monitoring networks, and to produce information graphics was written in the R programming language (R Development Core Team, 2013). Functions and data sets specific to this study were collected in an R package (a cross-platform extension of the R base system) called "ObsNetwork". Examples of how to use these functions and data sets are included in the package documentation (appendix A). In addition to the base packages included with R, ObsNetwork depends on the following contributed packages available on the Comprehensive R Archive Network (CRAN):

- *sp*: provides classes and methods for spatial data (Pebesma and Bivand, 2005; Bivand and others, 2008);

- *rgdal*: provides bindings to the Geospatial Data Abstraction Library (GDAL) and access to the cartographic projections library (PROJ.4) (Keitt and others, 2012);

- *raster*: performs geographic analysis and modeling with raster data (Hijmans and van Etten, 2012);

- *gstat*: performs spatial geostatistical modeling (Pebesma, 2004); and

- *GA*: implements genetic algorithms using a flexible general purpose set of tools (Scrucca, 2013).

ObsNetwork is in the public domain because it contains materials that originally came from the USGS. R and other package dependencies have more restrictive licenses. The code and documentation, including data sets used in this report, are available online for downloading from a software repository at https://github.com/jfisher-usgs/ObsNetwork. As this code is revised or updated, new versions will be made available for downloading from this site. Version information about R and required packages used in this report are shown in appendix B.

Computer Hardware

Genetic algorithms can be very demanding in terms of computer time; therefore, a brief summary of the computer hardware used in this study provides context for reported computation times (that is, the elapsed time for a GA run). The computer used for experimentation was equipped with a single Intel™ Xeon™ central processing unit (version X5687, four cores) at 3.6 gigahertz, and 32 gigabytes (GB) of random-access memory (RAM) at 1,333 megahertz. Four GB of RAM were allocated for each GA run; failures owing to memory restrictions were never a problem. GA runs often were made simultaneously (no more than four runs at a time) to leverage the multi-core processor; therefore, computation times should be viewed as approximate values.

Results and Discussion

The GA-based designs of water-level monitoring networks in the ESRP are based on two phases of sensitivity analysis. The first phase examines the relationships between select control parameters and optimal solutions and identifies control parameter values that optimize model performance. The second phase analyzes the trade-offs associated with changes in the number of wells to remove from the existing monitoring network (n_r). Parameter values for both phases of the sensitivity analysis are shown in table 1.

Model Performance

A series of GA runs were conducted for some of the control parameters (number of sites removed, kriging grid resolution, population size, elitism rate, crossover probability, and mutation probability) to better understand the sensitivity of the algorithm to incremental changes in the parameters (table 2, figs. 11 and 12), and to find reasonable settings for optimizing the existing monitoring networks. For example, a solution to the multi-objective function ("best fitness value") was determined for seven different population sizes while holding all other control parameters constant ("base-case values" in table 1). The functional relationship between best fitness value and population size provides a mechanism for evaluating the sensitivity of the GA. The analysis, however, ignores all interdependencies between the control parameters. Although the assumption that there are no interdependencies between the control parameters is not strictly valid, the analysis is believed to provide an adequate means for evaluating algorithm sensitivity. Four types of performance measures are considered throughout the analysis:

Table 1.   Parameter values used for base-case conditions and the final optimizations of the water-table monitoring networks of the eastern Snake River Plain aquifer, Idaho.

[**Control parameter**: parameters that control the optimization of the monitoring networks. **Base-case value**: parameter values specified as base-case conditions in the model performance phase of sensitivity. **Final value**: parameter values used in the final optimization of the monitoring networks]

Control parameter	Abbreviation	Unit	Base-case value	Final value
Kriging analysis				
Theoretical semivariogram (spherical model)				
Nugget	g	square meter	0.0	0.0
Sill	c	square meter	1,948.5	1,948.5
Range	r	meter	159,991.0	159,991.0
Spatial resolution of uniform kriging grid				
Length of square block side	ℓ	kilometer	2.5	[1]2.5 / [2]1.5
Multi-objective problem formulation				
Number of well sites to remove from the existing monitoring network	n_r	unitless	40	10, 20, 40, 60, 80
Weighting coefficients on individual objective functions				
Weight on the standard error function (f_1)	w_1	unitless	100	100
Weight on the root-mean-square error function (f_2)	w_2	unitless	1	1
Weight on the mean standard deviation function (f_3)	w_3	unitless	1	1
Weight on the mean measurement error function (f_4)	w_4	unitless	1	1
Penalty function				
Penalty coefficient	C	meter	1,000,000	1,000,000
Genetic algorithm				
Population size	n_{pop}	unitless	2,000	2,000
Genetic operations				
Elitism rate	r_e	unitless	0.05 (5 percent)	0.05 (5 percent)
Crossover probability	p_c	unitless	0.80 (80 percent)	0.80 (80 percent)
Mutation probability	p_m	unitless	0.05 (5 percent)	0.30 (30 percent)
Maximum number of times a child chromosome can be aborted during crossover	n_a	unitless	10	10
Terminating conditions				
Maximum number of iterations	n_{iter}	unitless	100	infinity
Maximum number of consecutive iterations without any improvement in the best fitness value	n_{run}	unitless	infinity	50

[1] Specified for the Federal-State Cooperative water-level monitoring network (Co-op network).

[2] Specified for the U.S. Geological Survey-Idaho National Laboratory water-level monitoring network (USGS-INL network).

1. Best fitness value (F', ideally near the minimum), in meters.

2. Computation time (reasonable given computational resources and ideally small), in hours.

3. The percentage of chromosomes that invoke the penalty function (equation 35), where GA performance decreases as the number of calls to the penalty function increases.

4. Number of consecutive iterations without any improvement in the best fitness value (n_{run}). The magnitude of value n_{run} gives some indication as to whether the global solution was found. That is, the probability of a GA solution being globally optimal increases as the number of consecutive iterations without improvement increases.

Table 2. Sensitivity of the genetic algorithm to incremental changes in the control parameters, eastern Snake River Plain, Idaho.

[**Control parameter:** parameters that control the optimization of the water-level monitoring network. A **bolded** control parameter value indicates base-case conditions (see table 1). **Number of sites removed:** the number of well sites to remove from the existing monitoring network (n_r). **Kriging grid resolution:** the spatial resolution of the uniform kriging grid described using the length of a grid block side (ℓ). **Population size:** the number of chromosomes in a population (n_{pop}). **Elitism rate:** the fraction of the population that is guaranteed to survive to the next iteration (r_e). **Crossover probability:** the probability that crossover will occur between two parent chromosomes (p_c). **Mutation probability:** the probability that a mutation will occur on a child chromosome (p_m). **Best fitness value:** the smallest fitness value (F'). **Computation time:** the time required to run the genetic algorithm. **Percent penalty:** the percentage of chromosomes that invoke the penalty function. **Number of times best-fitness repeated:** the number of consecutive iterations without any improvement in the best-fitness value. **Abbreviations:** Co-op network, Federal-State Cooperative water-level monitoring network; USGS-INL network, U.S. Geological Survey-Idaho National Laboratory water-level monitoring network; km, kilometer; m, meter; h, hour]

Control parameter		Co-op network				USGS-INL network			
Name	Value	Best fitness value (m)	Computation time (h)	Percent penalty	Number of times best fitness repeated	Best fitness value (m)	Computation time (h)	Percent penalty	Number of times best fitness repeated
Number of sites removed	10	1,374.315	14.8	0.0	78	1,174.581	8.8	0.1	67
	20	1,378.496	16.1	0.2	30	1,174.913	9.7	0.1	39
	40	1,390.061	16.7	0.4	2	1,175.230	11.7	0.5	10
	60	1,409.370	15.4	0.8	2	1,175.716	10.9	0.8	1
	80	1,443.009	15.5	1.0	4	1,177.684	10.9	1.4	2
Kriging grid resolution (km)	1.5	1,382.731	38.2	0.1	4	1,145.589	24.2	1.5	1
	2.0	1,386.504	22.7	0.2	9	1,166.779	15.2	0.3	6
	2.5	1,390.061	16.7	0.4	2	1,175.230	11.7	0.5	10
	3.0	1,396.676	11.6	0.2	1	1,186.873	8.0	0.3	3
	3.5	1,400.436	9.0	0.3	27	1,206.887	6.4	1.1	5
	4.0	1,407.150	8.1	0.2	1	1,218.568	5.4	0.2	20
Population size	500	1,391.565	3.9	0.1	2	1,175.293	2.8	0.1	5
	1,000	1,390.643	7.7	0.3	5	1,175.230	5.3	1.2	1
	1,500	1,390.008	10.8	0.5	18	1,175.249	7.5	0.5	1
	2,000	1,390.061	16.7	0.4	2	1,175.230	11.7	0.5	10
	3,000	1,389.979	22.6	0.5	10	1,175.231	15.9	0.3	2
	4,000	1,389.848	30.6	0.3	1	1,175.223	20.1	2.3	13
	5,000	1,389.848	37.4	0.5	11	1,175.223	25.9	1.1	17
Elitism rate	0.01	1,390.488	15.0	0.1	1	1,175.247	9.3	0.6	2
	0.03	1,390.243	15.4	0.3	4	1,175.234	10.1	1.0	3
	0.05	1,390.061	16.7	0.4	2	1,175.230	11.7	0.5	10
	0.10	1,389.958	15.0	1.3	2	1,175.239	10.9	0.8	1
	0.15	1,390.032	15.4	0.3	5	1,175.250	11.0	0.3	3
	0.20	1,390.066	15.1	0.5	5	1,175.234	10.7	2.4	4
	0.30	1,390.026	14.3	0.3	4	1,175.229	9.5	0.7	5
	0.40	1,389.955	13.6	0.3	2	1,175.242	8.5	1.1	7
	0.50	1,390.293	10.3	0.4	3	1,175.252	7.1	1.1	1
Crossover probability	0.5	1,390.137	10.7	0.1	3	1,175.246	7.6	0.2	3
	0.6	1,390.126	12.0	0.2	7	1,175.256	8.2	1.1	1
	0.7	1,390.227	14.3	0.8	29	1,175.236	9.4	1.0	1
	0.8	1,390.061	16.7	0.4	2	1,175.230	11.7	0.5	10
	0.9	1,390.022	18.8	0.1	6	1,175.244	11.6	0.7	1
	1.0	1,390.327	20.2	0.4	33	1,175.224	13.5	0.2	12
Mutation probability	0.005	1,390.957	16.6	0.1	6	1,175.272	11.6	0.1	4
	0.020	1,390.747	16.5	0.2	3	1,175.259	9.1	0.2	5
	0.050	1,390.061	16.7	0.4	2	1,175.230	11.7	0.5	10
	0.100	1,390.041	16.7	0.4	12	1,175.225	11.8	0.5	1
	0.200	1,389.848	17.0	1.0	2	1,175.228	12.0	1.3	2
	0.300	1,389.870	17.6	2.3	7	1,175.223	10.4	3.9	1
	0.400	1,389.848	16.5	2.1	2	1,175.224	11.6	2.2	5
	0.500	1,390.026	16.9	3.0	15	1,175.224	11.2	8.1	1

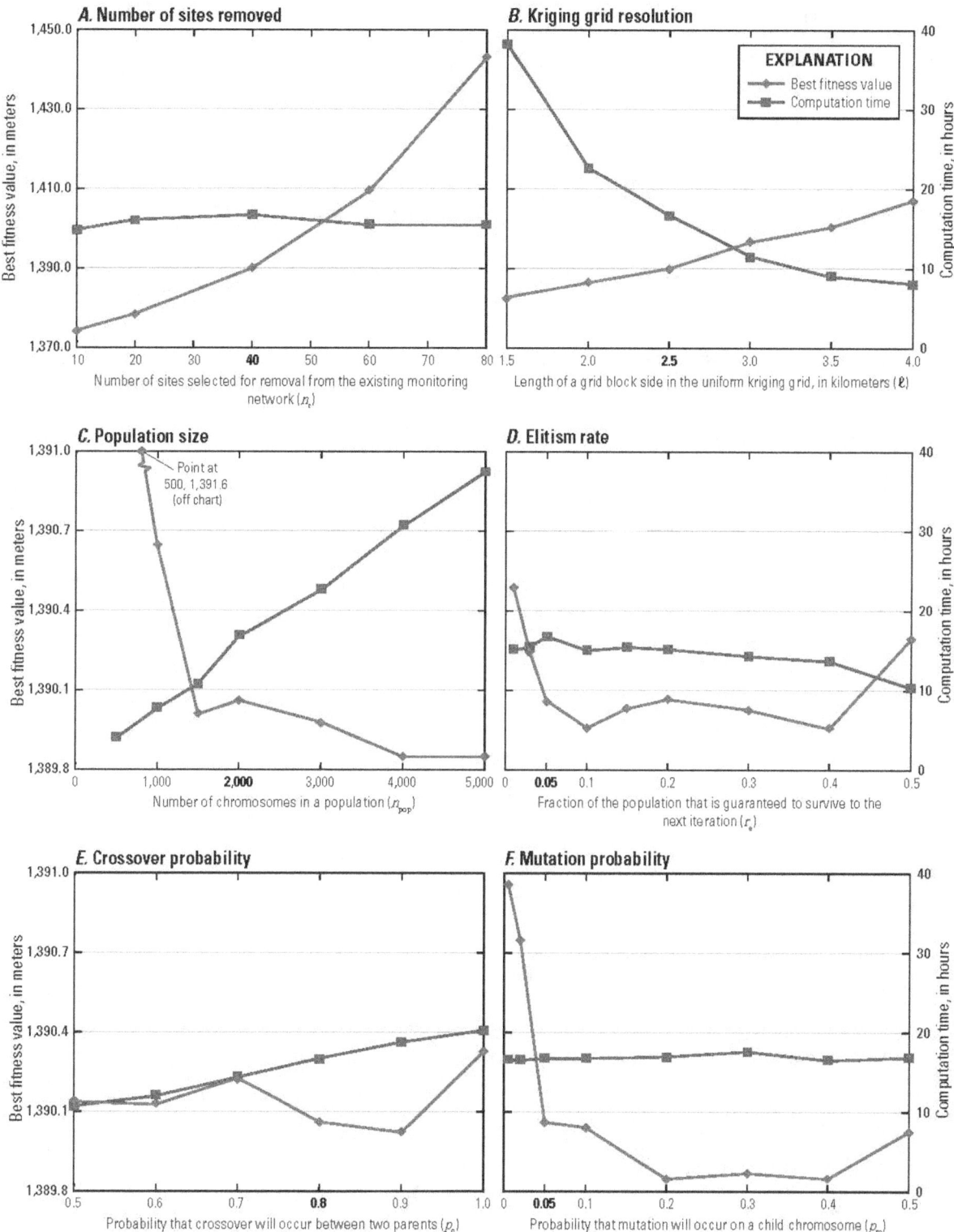

Figure 11. Sensitivity of the best fitness value and computational time to changes in the (*A*) number of sites removed, (*B*) kriging grid resolution, (*C*) population size, (*D*) elitism rate, (*E*) crossover probability, and (*F*) mutation probability, Federal-State Cooperative water-level monitoring network, eastern Snake River Plain, Idaho. A bold control parameter value indicates base-case conditions (see table 1).

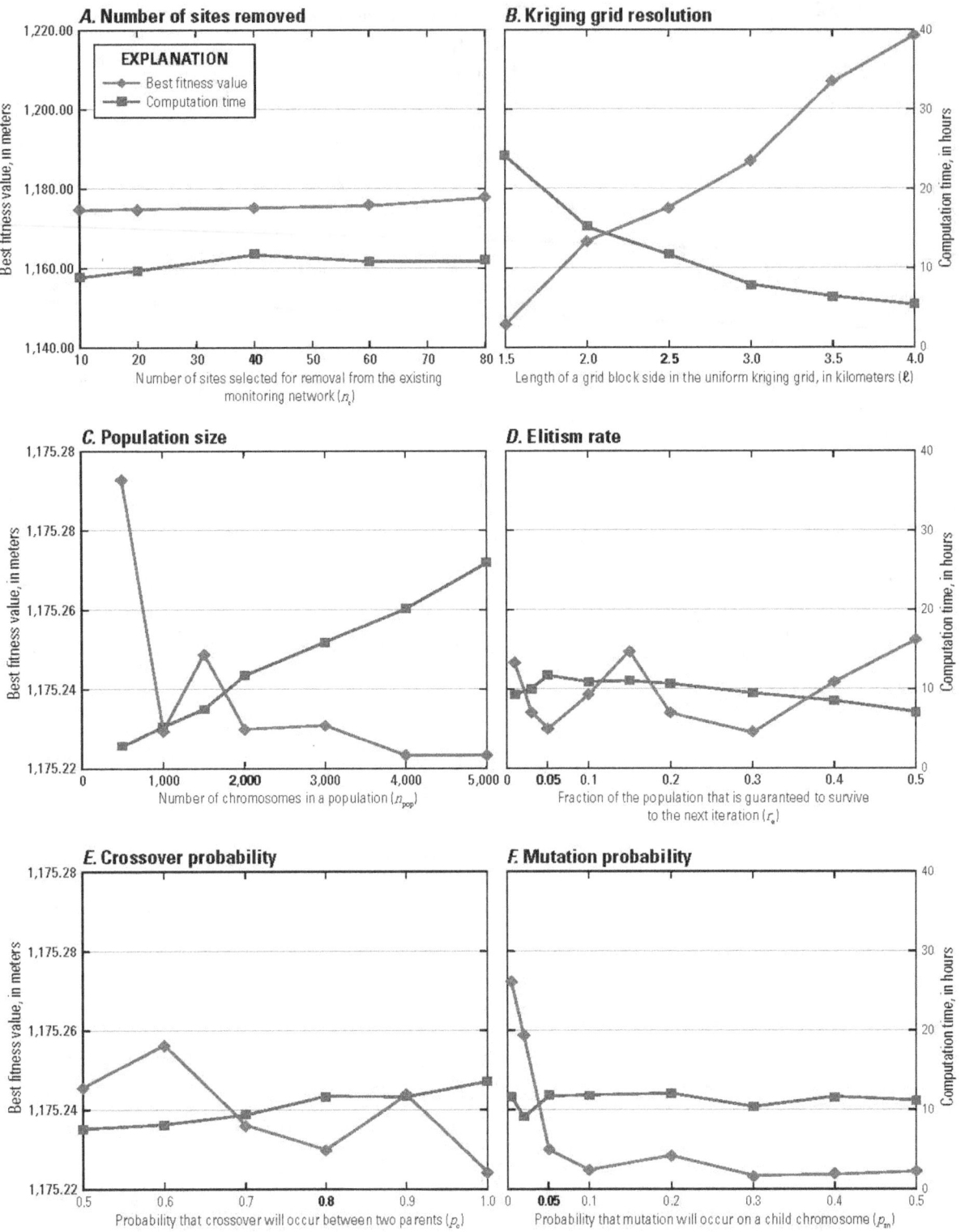

Figure 12. Sensitivity of the best fitness value and computational time to changes in the (*A*) number of sites removed, (*B*) kriging grid resolution, (*C*) population size, (*D*) elitism rate, (*E*) crossover probability, and (*F*) mutation probability, U.S. Geological Survey-Idaho National Laboratory water-level monitoring network, eastern Snake River Plain, Idaho. A bold control parameter value indicates base-case conditions (see table 1).

The GA sensitivity analysis includes runs for the Co-op and USGS-INL networks, and examines the four performance measures and the tradeoff between these measures as a function of the control parameters (table 2). The response curves for best fitness and computation time are shown for all GA runs in figures 11 (Co-op network runs) and 12 (USGS-INL network runs). Performance is considered best when the best fitness value is near the minimum value and before computation time becomes too long.

Results of the sensitivity analysis pertaining to each of the control parameters are summarized in the six subsections that follow:

Number of Sites Removed

The number of sites removed from an existing monitoring network (n_r) has a significant effect on the best fitness value because of the strong dependence of mean standard error (f_1) and root-mean-square error (f_2) on n_r. Increasing n_r results in increased values of f_1 and f_2, which, in turn, increases the fitness value (weighted sum of objectives, equation 29). A rapid worsening of best fitness value starting between 20 and 40 sites is shown in figure 11A, indicating that the removal of less than 20 sites has a relatively small effect on the ability of the network to represent the water table. A rapid worsening between 60 and 80 sites is shown in figure 12A, indicating that the removal of less than 60 sites has a relatively small effect on the ability of the network to represent the water table. By comparison, these observations imply that 20 sites can be removed from the Co-op network with a relatively small degradation of the estimated water-table map, and 60 sites can be removed from the USGS-INL network before the water-table map degradation increases more rapidly. This is not unexpected given the high network density around site facilities in the USGS-INL network (fig. 2). The possible redundancy between water-level measurements in these data clusters is quite large; therefore, many of these wells can be removed with-little to-no effect on the estimated water-table map. Kriging compensates for the effects of data clustering by treating clusters more like single points (Isaaks and Srivastava, 1989, p. 300).

The effect on computation time from incremental changes in n_r was small (less than 2 hours). Peaking at 40 sites for both monitoring networks is shown in figures 11A and 12A, indicating a non-linear relationship between n_r and computation time. This nonlinearity is attributed to the computational cost of kriging; the search for an optimal network requires many simulations of the water-table map, making kriging the most expensive operation in the optimization problem. Given that water levels are estimated at the locations of removed wells ($\hat{z}$ in equation 32), a reduction in n_r requires fewer estimates, which, in turn, decreases the computation time for kriging. The cost of kriging also depends on the number of wells in the reduced monitoring network (n in equation 4), where an increase in n_r results in a smaller network size and requires less computation time for kriging when compared to the costs associated with kriging a larger network. Finally, the nonlinearity in the computation time response curves results from the interplay between these two opposing forces (the size of the reduced monitoring network increases as the number of wells removed from the existing network decreases and the reverse), and any cost savings associated with a reduced number of estimates or smaller network size is cancelled out when n_r is equal to 40. The number of times the penalty function (equation 35) is invoked can also affect the computation time; increasing the number of penalty calls results in a decrease in computation time because the penalty function does not require kriging. The percentage of chromosomes that invoke the penalty function was 1.0 percent or less for the Co-op network and 1.4 percent or less for the USGS-INL network indicating that computation time was unaffected by penalty calls (table 2).

The number of iterations needed to converge on a solution increased as the number of removed wells increased, as indicated by the increase in the number of consecutive iterations without any improvement in the best-fitness when n_r was decreased (table 2). This should be expected given that the number of possible combinations increases as n_r is increased. For example, in the Co-op network there are 3.3×10^{15} possible network configurations when n_r equals 10 and 5.2×10^{48} configurations when n_r equals 80.

Kriging Grid Resolution

Refining the spatial resolution of the uniform kriging grid (measured as the length of a grid block side, ℓ) linearly decreased the mean standard error (f_1) (and consequently the best fitness value) (figs. 11B and 12B). Because the standard error (σ_{UK}) is calculated at the nodes of the kriging grid, a finer grid resolution describes in more detail the depressions in the standard error surface near the observation wells (where $\sigma_{UK} = 0$ at a well site). This added detail increases the positive skew in the standard error probability distribution, which implies that the mean value is decreased. Linear regression applied to the best fitness values indicates that the USGS-INL network (slope = 28.4 m/km, $R^2 = 0.987$) is much more sensitive to changes in grid resolution than the Co-op network (slope = 9.7 m/km, $R^2 = 0.989$). This suggests that the relatively dense resolution of the USGS-INL network supports a finer resolution of the kriging grid. An exponential increase in computation time for refinements in the kriging grid resolution occurred as accuracy of the mean standard error from refinements in grid resolution was improved (figs. 11B and 12B). An ℓ equal to 2.5 km for the Co-op network and 1.5 km for the USGS-INL network provides the optimal tradeoff between best fitness value and computation times.

Population Size

Population size is the number of chromosomes in a population (n_{pop}). If the population size is too small, the algorithm may explore too little of the solution space to find a suitable solution (Marczyk, 2004). A rapid worsening of the best fitness values starting at between 1,000 and 1,500 chromosomes is shown in figure 11C, indicating that a suitable GA solution was not available for population sizes of less than 1,500. A rapid decrease in best fitness values starting at between 500 and 1,000 chromosomes, and again between 1,500 and 2,000 chromosomes, is shown in figure 12C, indicating that a population size of less than 2,000 is susceptible to suboptimal solutions. By comparison, this implies that the minimum population size needed to optimize the Co-op and USGS-INL networks is 1,500 and 2,000, respectively. Increasing the population size enlarges the search space, which, in turn, increases the computation time. The computation time (figs. 11C and 12C) increases linearly at a rate of about 3.8 ($R^2 = 0.997$) and 2.6 ($R^2 = 0.995$) hours, respectively, for every 500 chromosomes added to the population. The fitness response to changes in population size for both networks indicates that a population must be composed of at least 2,000 chromosomes to find a good solution while minimizing computation time.

Elitism Rate

Elitism is the fraction of the population that is guaranteed to survive to the next iteration of the GA. This subset of chromosomes preserves the best solutions from one iteration to the next. Elitism can be an effective method for improving the efficiency of the algorithm; however, if set too large, it can decrease genetic diversity and potentially result in the global solution being overlooked (that is, it converges to a local minimum). A rapid worsening of the best fitness values starting at between 0.05 (5 percent) and 0.10 (10 percent) is shown in figure 11D, indicating a performance loss for elitism rates of less than 0.10; a decrease in elitism slows convergence. Another rapid worsening of the best fitness values was observed between 0.4 and 0.5, indicating premature convergence for elitism rates of greater than 0.4. A worsening of the best fitness values at the extremes (less than 0.05 and greater than 0.3) also is shown in figure 12D, however, a peak in the fitness at 0.15 may indicate that increasing elitism rates beyond 0.05 may result in a significant loss in genetic diversity.

A decrease in the computation time for increased elitism rate was observed in both networks for values between 0.2 and 0.5 (figs. 11D and 12D). Elitist chromosomes survive to the next iteration along with their fitness value, and non-elitist

chromosomes calculate a fitness value only when they are derived from crossover operations (80 percent chance), mutation operations (5 percent chance), or both. The fitness calculation is computationally expensive (that is, unless the penalty function is invoked); therefore, an increase in the fraction of the population that does not require a fitness calculation results in a shorter computation time. For elitist rates of greater than 0.2, this fraction is guaranteed to increase; whereas, for elitism rates of less than 0.2, this fraction is a function of the elitism rate, the crossover probability, and the mutation probability. Computational cost savings associated with a large elitism rate are relatively insignificant when compared to the potential loss in genetic diversity; therefore, an elitism rate of 0.05 (5 percent) is used to facilitate the search for a globally optimal solution.

Crossover Probability

The crossover probability controls the rate at which solutions are subjected to crossover. The larger the value of crossover probability, the quicker potential new solutions are introduced into the population. If crossover probability is too large, chromosomes with good fitness values are discarded faster than selection can exploit them. However, if crossover probability is too small, the search may stagnate owing to the smaller exploration rate. As expected, the best fitness value generally improves for increased values of crossover probability, as shown in figure 11E. The exception to this downward trend is a rapid worsening of the best fitness value between 0.9 (90 percent) and 1.0 (100 percent), indicating that crossover probabilities of greater than 0.9 may be too large. A general downward trend also is shown in figure 12E; however, relatively large best fitness values at 0.6 and 0.9 indicate a significant amount of uncertainty associated with this trend. For both networks, an almost-linear increase in computation time was observed for increasing values of crossover probability (figs. 11E and 12E). Increasing the crossover probability increases the fraction of the population requiring fitness calculations, which, in turn, increases the computation time. A comparison of the tradeoff between best fitness values and computation times indicates that, for both networks, selection of a 0.8 (80 percent) crossover probability will give relatively accurate estimates of the optimal monitoring network.

Mutation Probability

Mutation is used to introduce genetic diversity between iterations and to prevent convergence on a local minimum (sub-optimal solution). If mutation probability is too large, the GA will have difficulty converging on a suitable solution and the search becomes random. However, if the mutation probability is too small, the algorithm's ability to explore the solution space will be greatly diminished. That is, the population of chromosomes becomes so similar that evolution slows or even stops. A worsening of the best fitness value between 0.005 (0.5 percent) and 0.2 (20 percent) is shown in figure 11F, indicating that mutation probabilities of less than 0.2 are too small. The fitness again worsens between 0.4 and 0.5, indicating that mutation probabilities of greater than 0.4 are too large. The worst best fitness values are shown in figure 12F between 0.3 and 0.005, indicating that mutation probabilities of less than 0.3 are too small. Changes in computation time are negligible for both networks (less than 1.1 hours for the Co-op network and less than 3.0 hours in the USGS-INL network; figs. 11F and 12F). This is not unexpected given that the fraction of the population requiring fitness calculations is determined primarily by the elitism rate (20 percent) and crossover probability (80 percent). A mutation probability value of 0.3 (30 percent) provides an excellent starting value for further analyses.

Numerical results indicate similar responses between networks; however, best fitness values and computation times are much larger for GA runs conducted with the Co-op network. For example, GA solutions determined using base-case conditions indicate a fitness and computation time for the Co-op network that are, respectively, 214.8 m larger and 5.0 hours longer than for the USGS-INL network. The increased magnitude of best fitness values is attributed to a network resolution that is relatively low for the Co-op network when compared to the USGS-INL network, where lower network resolutions typically increase values of mean standard error (f_1) and root-mean-square error (f_2), which, in turn increase the fitness value. The longer computation times are owing to the relatively large number of nodes in the kriging grid of the Co-op network $(n_n = 4,690)$ when compared to the USGS-INL network $(n_n = 774)$. Mean standard error (equation 31) requires estimates of standard error (σ_{UK}) at each node in the kriging grid; therefore, as the total number of standard error estimates is increased, so too is the problem size (together with the computation time) increased.

The set of control parameter values identified as optimizing model performance and used in the final optimizations of the existing water-level monitoring networks is given in table 1. The most significant changes from base-case conditions are (1) a reduction in the kriging grid resolution for the USGS-INL network from 2.5 to 1.5 m, (2) an increase in mutation probability from 0.05 to 0.30, and (3) the implementation of a new stopping criterion. For the

final optimizations, the GA terminates only after exceeding 50 consecutive iterations without any improvement in the best fitness value, thereby facilitating the search for a globally optimal solution.

Weighting Coefficients

For multi-objective problems, identifying a single solution that simultaneously minimizes each individual objective function (equations 31–34) is almost impossible. That is, any single individual objective value often can be improved only by degrading at least one of the other objective values. Combining the individual objective functions in a single weighted-objective function is subjective, requiring that a decision maker provide the weights. The weighted multi-objective function also is ill-suited for determining tradeoffs among individual objective functions; therefore, the algorithm's sensitivity to changes in weights is not examined. Because the objective functions are simply weighted and added to produce a single fitness value, the function with the largest range dominates GA evolution. A poor value for the objective function with the larger range degrades the overall fitness much more than a poor value for the function with the smaller range (Bentley and Wakefield, 1997).

For this study, an emphasis was placed on the estimation uncertainty design criterion (f_1) by setting w_1 equal to 100, and w_2, w_3, and w_4 equal to 1. This assumes that network coverage is more important than the other design criteria: preserving localized features in the water table (f_2), maintaining temporal variations in water-level measurements (f_3), and reducing measurement error (f_4). The range of weighted-objective values in solution space indicates the relative influence of each design criterion in determining the optimal solution. For a given GA run and design criterion, a weighted-objective value is calculated for every evaluation of the weighted sum objective function (F in equation 29); the range of this value is defined as the difference between the largest and smallest of its individual component values. The range of each weighted-objective function is given in table 3 for GA runs based on both networks and changing the number of wells to remove from the original network (n_r). As indicated by their ranges, the relative influence of each design criterion on the solution is given in decreasing order of importance as f_1, f_2, f_3, and f_4. As intended, design criteria f_3 and f_4 have little control over GA evolution and only after f_1 and f_2 have been minimized to their fullest possible extent. The criteria f_3 and f_4 were assumed much less important than f_1 and f_2.

Table 3. Range of weighted-objective values in solution space (the collection of all possible solutions to the optimization problem) for changes in the number of sites to remove, eastern Snake River Plain, Idaho.

[**Number of sites removed:** well sites to remove from the existing monitoring network (n_r). **Weighted objective function:** the individual objective function multiplied by its weighting coefficient. **Minimum, Maximum, and Range:** the minimum, maximum, and range of all calculated weighted objective values during the genetic algorithm search, respectively. Entry in **bold** indicates the weighted objective function with the largest range. **Control parameter values:** a kriging grid resolution of 2.5 kilometers for the Co-op network and 1.5 kilometers for the USGS-INL network, population size of 2,000, elitism rate of 0.05, crossover probability of 0.80, mutation probability of 0.30, and terminates after 50 consecutive iterations without any improvement in the best fitness value. **Abbreviations:** Co-op network, Federal-State Cooperative water-level monitoring network; USGS-INL network, U.S. Geological Survey-Idaho National Laboratory water-level monitoring network; m, meter; w, weighting coefficient; f, individual objective function]

Number of sites removed	Weighted objective function	Co-op network			USGS-INL network		
		Minimum (m)	Maximum (m)	Range (m)	Minimum (m)	Maximum (m)	Range (m)
10	$w_1 f_1$	1,371.00	1,500.18	**129.14**	1,143.79	1,252.90	**109.11**
	$w_2 f_2$	0.70	105.48	104.73	0.06	4.42	4.36
	$w_3 f_3$	0.70	4.56	3.90	0.64	2.55	1.90
	$w_4 f_4$	0.60	0.71	0.11	0.02	0.06	0.05
20	$w_1 f_1$	1,374.13	1,528.82	**154.69**	1,143.80	1,307.87	**164.07**
	$w_2 f_2$	1.17	93.17	92.00	0.10	3.34	3.24
	$w_3 f_3$	0.97	3.77	2.80	1.06	2.12	1.06
	$w_4 f_4$	0.56	0.76	0.19	0.01	0.07	0.06
40	$w_1 f_1$	1,383.15	1,620.94	**237.79**	1,143.86	1,484.57	**340.71**
	$w_2 f_2$	3.15	74.11	70.96	0.14	2.67	2.53
	$w_3 f_3$	1.12	3.08	1.95	1.29	1.96	0.67
	$w_4 f_4$	0.51	0.83	0.32	0.01	0.08	0.07
60	$w_1 f_1$	1,400.00	1,788.06	**388.06**	1,144.22	1,606.08	**461.85**
	$w_2 f_2$	5.43	67.97	62.54	0.20	3.03	2.83
	$w_3 f_3$	1.40	2.63	1.23	1.38	1.89	0.51
	$w_4 f_4$	0.48	0.89	0.42	0.01	0.09	0.08
80	$w_1 f_1$	1,429.98	2,046.18	**616.21**	1,146.25	1,653.73	**507.48**
	$w_2 f_2$	6.11	77.89	71.78	0.28	2.61	2.33
	$w_3 f_3$	1.50	2.63	1.13	1.40	1.81	0.41
	$w_4 f_4$	0.43	0.95	0.53	0.01	0.11	0.10

Optimized Monitoring Networks

Each water-level monitoring network was optimized five times: removing 10, 20, 40, 60, and 80 observation wells from the original network (see "Final values" in table 1 for control parameter values). Wells identified for removal are shown in table 6 (at back of report). Here, "Times identified" indicates the number of times the observation well was identified for removal in the five GA runs for each network. For example, 6 of the 10 wells selected for removal from the Co-op network were identified in all five GA runs (wells 8, 80, 120, 124, 140, and 164), and 4 of the 10 wells were identified in four of the GA runs (wells 34, 84, 91, and 146). By comparison, 8 of the

10 wells selected for removal from the USGS-INL network were identified in all five GA runs (wells 184, 187, 188, 189, 213, 245, 314, and 321), 1 of the 10 wells was identified in four of the GA runs (well 206, in all but the GA run removing 80 wells), and 1 of the 10 wells was identified only once (well 251). Each GA run provides a unique solution that is entirely dependent on the number of wells to remove from the original network. That is, the solutions are non-sequential; wells identified for removal in the GA run removing 10 wells are not required to be part of the solution for the GA run removing 20 wells. The relatively large values of times identified, however, indicate that a consistent group of wells provides little-to-no beneficial added information.

A summary of the GA runs is shown in table 4. Performance measures for each GA run include: best fitness value, number of iterations, computation time, root-mean-square deviation (RMSD), and the percent local error (PLE). The RMSD (ideally small) is a measure of the difference between kriged water-table maps estimated from measurements in the original and optimized networks, and expressed as:

$$RMSD = \sqrt{\frac{\sum_{i=1}^{n_n} \left[\hat{z}_{orig}(s_{n,i}) - \hat{z}(s_{n,i}) \right]^2}{n_n}} \quad (36)$$

where

n_n is the number of nodes in the kriging grid;

$s_{n,i}$ is the spatial coordinate of node i in the kriging grid;

$\hat{z}_{orig}$ is the estimate of the water-level elevation based on the original network, in meters; and

$\hat{z}$ is the estimate of the water-level elevation based on the reduced network, in meters.

The PLE (ideally small) is the maximum error introduced by removing n_r wells, divided by the relief across the kriged water-table map based on the original network, and expressed as:

$$PLE = \frac{100 \cdot maximum\left[\left| \hat{z}_{orig}(s_{n,i}) - \hat{z}(s_{n,i}) \right| \right]}{maximum\left[\hat{z}_{orig}(s_{n,i}) \right] - minimum\left[\hat{z}_{orig}(s_{n,i}) \right]} \quad (37)$$

for $i = 1, ..., n_n$.

A comparison with the Co-op network GA runs implemented using base-case control parameter values indicates large gains in model performance using the final values optimized for model performance (table 1). Subtracting best fitness values calculated using the optimized (final) control parameters (table 4) from values calculated using base-case control parameters (table 2) gives the change in best fitness. Best fitness values decreased by 0.016, 0.191, 0.407, and 2.644 m for GA runs removing 20, 40, 60, and 80 wells, respectively. No change was observed for GA runs removing 10 wells, indicating this run has a relatively rapid rate of convergence on the optimal solution. The solution space for a GA run is proportional to the number of wells removed; therefore, as n_r increases, so, too, do the number of iterations needed to converge on an optimal solution increase (together with computation time). This was the case for all runs except the Co-op network GA run removing 60 wells (table 4). Random sampling in the GA may explain this data abnormality.

Table 4. Genetic algorithm searches summarized for optimized water-level monitoring networks, eastern Snake River Plain, Idaho.

[**Number of sites removed:** well sites removed from an existing monitoring network (n_r). **Best fitness value:** the smallest fitness value (F'). **Number of iterations:** the number of completed genetic algorighm (GA) iterations. **Computation time:** the time required to run the GA. RMSD: root-mean-square deviation, a measure of the difference between kriged water-table surfaces estimated using the optimized reduced network and the existing network. **Percent local error:** between water-table maps estimated using the existing and optimized reduced networks. **Control parameter values:** a kriging grid resolution of 2.5 kilometers for the Co-op network and 1.5 kilometers for the USGS-INL network, population size of 2,000, elitism rate of 0.05, crossover probability of 0.80, mutation probability of 0.30, and terminates after 50 consecutive iterations without any improvement in the best fitness. **Abbreviations:** Co-op network, Federal-State Cooperative water-level monitoring network; USGS-INL network, U.S. Geological Survey-Idaho National Laboratory water-level monitoring network; m, meter; h, hour]

Number of sites removed	Co-op network					USGS-INL network				
	Best fitness value (m)	Number of iterations	Computation time (h)	RMSD (m)	Percent local error	Best fitness value (m)	Number of iterations	Computation time (h)	RMSD (m)	Percent local error
10	1,374.315	86	11.8	0.061	0.24	1,144.979	81	16.0	0.002	0.23
20	1,378.480	108	17.6	0.143	0.39	1,145.279	131	37.6	0.005	0.30
40	1,389.870	181	28.6	0.598	1.10	1,145.586	187	55.1	0.006	0.30
60	1,408.964	245	39.5	1.276	2.92	1,146.171	203	53.1	0.019	1.02
80	1,440.365	197	30.8	1.736	2.94	1,148.343	246	64.8	0.051	1.51

The response curves for best-fitness, RMSD, and PLE are shown for all GA runs in figure 13. A rapid worsening of best fitness value, RMSD, and PLE, respectively, starting at between 20 and 40 sites, is shown in figures 13A–C, indicating that the removal of less than 20 wells has a relatively small effect on the ability of the network to represent the water table. A rapid worsening of best fitness value, RMSD, and PLE, respectively, starting at between 40 and 60 sites, is shown in figures 13D–F, indicating that the removal of less than 40 wells has a relatively small effect on the ability of the network to represent the water table. By comparison, this implies that 20 wells can be removed from the Co-op network with a relatively small degradation of the estimated water-table map, and 40 wells can be removed from the USGS-INL network before the water-table map degradation increases rapidly.

The spatial distribution of the difference between the two estimated surfaces (that is, the water table estimated from data in the original network minus the water table estimated from data in the reduced network) is shown for 10, 20, 40, 60, and 80 wells removed from the Co-op network (fig. 14) and the USGS-INL network (fig. 15). Relatively small differences with a spatial extent limited to near the removed well sites are shown in figures 14A and 14B. This is not unexpected given that the analysis of break-points indicates 20 wells can be removed from the Co-op network with relatively small degradation of the estimated water-table map. Wells primarily were removed from areas of high network density located in the northeast and south-central parts of the ESRP.

A high network density indicates that the redundancy between water-level measurements in these areas can be quite large; therefore, many of the wells in these areas can be removed with little-to no-effect on the estimated water-table map. Progressively larger differences with increasing spatial extent are shown in figures 14C–E. The areas excluded from well removal coincide with areas of rapid change in the water-table elevation and areas of the monitoring network that are sparsely populated with wells. For example, only a small number of wells were removed from areas of steeper hydraulic gradients along the margin of the ESRP and near Mud Lake (figs. 2 and 14E). Wells were never removed from the sparsely populated area in the west-central part of the ESRP; removing wells in these areas would significantly degrade the estimated water-table map.

For 10, 20 and 40 wells removed from the USGS-INL network, the magnitude and spatial extent of differences is negligible (figs. 15A–C) when compared to the distribution of corresponding Co-op network differences (figs. 14A–C). This is not unexpected given that the analysis of break-points indicates 40 wells can be removed from the USGS-INL network with relatively small degradation of the estimated water-table map. Wells were removed exclusively from areas of high network density near site facilities in the INL (fig. 2). Progressively larger differences with increasing spatial extent are shown in figures 15D and 15E. Removed wells were exclusively from areas of high network density near site facilities, indicating widespread data redundancy in this network.

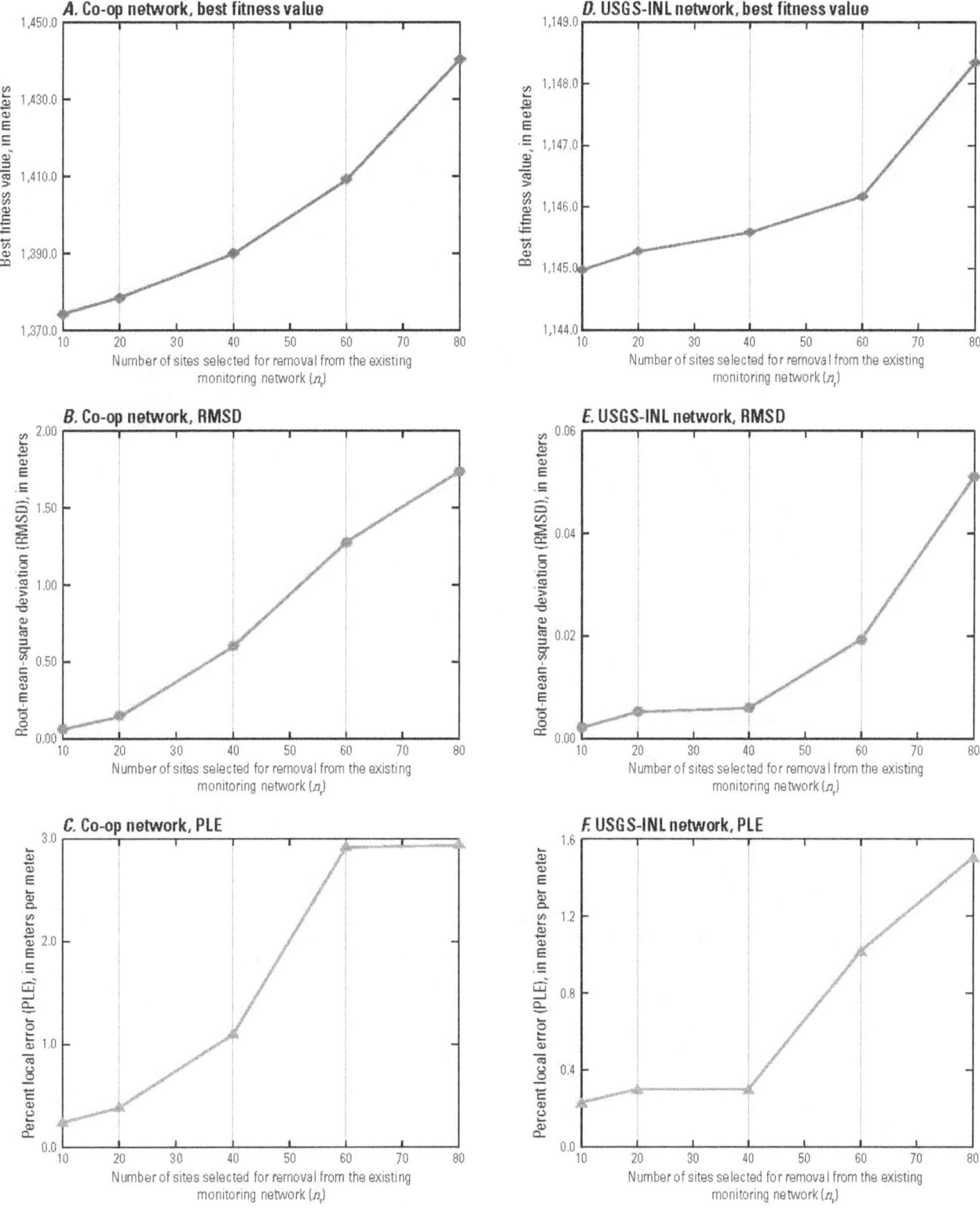

Figure 13.   Sensitivity of the best fitness value, root-mean-square deviation, and percent local error to changes in the number of sites removed from the existing (*A–C*) Federal-State Cooperative and (*D–F*) U.S. Geological Survey-Idaho National Laboratory water-level monitoring networks, eastern Snake River Plain, Idaho.

A. 10 wells removed

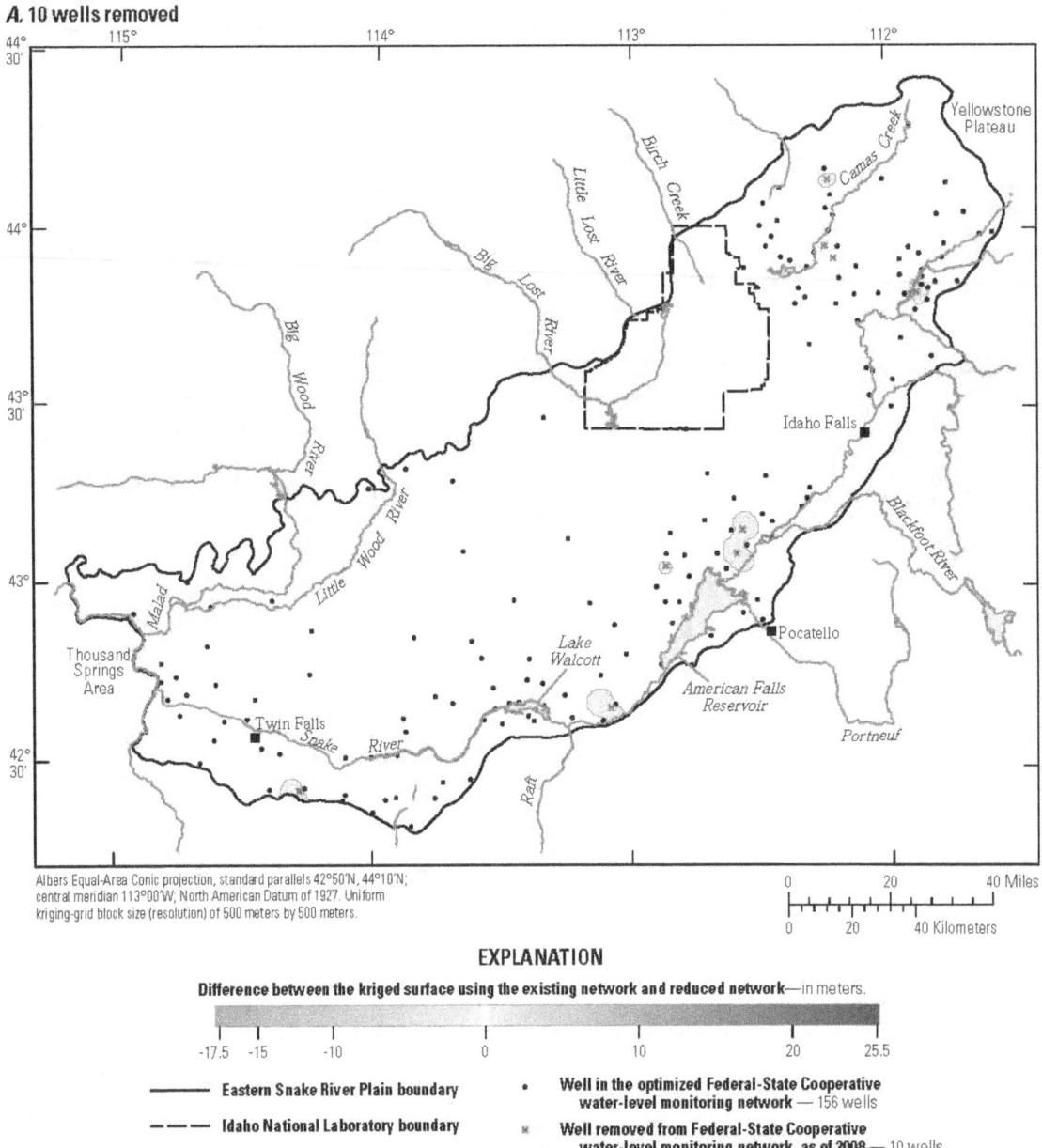

Albers Equal-Area Conic projection, standard parallels 42°50'N, 44°10'N;
central meridian 113°00'W, North American Datum of 1927. Uniform
kriging-grid block size (resolution) of 500 meters by 500 meters.

EXPLANATION

Difference between the kriged surface using the existing network and reduced network—in meters.

-17.5 -15 -10 0 10 20 25.5

——— Eastern Snake River Plain boundary

– – – Idaho National Laboratory boundary

• Well in the optimized Federal-State Cooperative
water-level monitoring network — 156 wells

✖ Well removed from Federal-State Cooperative
water-level monitoring network, as of 2008 — 10 wells

Figure 14. Difference between kriged water-table surfaces using the existing and reduced Federal-State Cooperative water-level monitoring network, after removing (*A*) 10, (*B*) 20, (*C*) 40, (*D*) 60, and (*E*) 80 optimally selected wells, eastern Snake River Plain, Idaho.

B. 20 wells removed

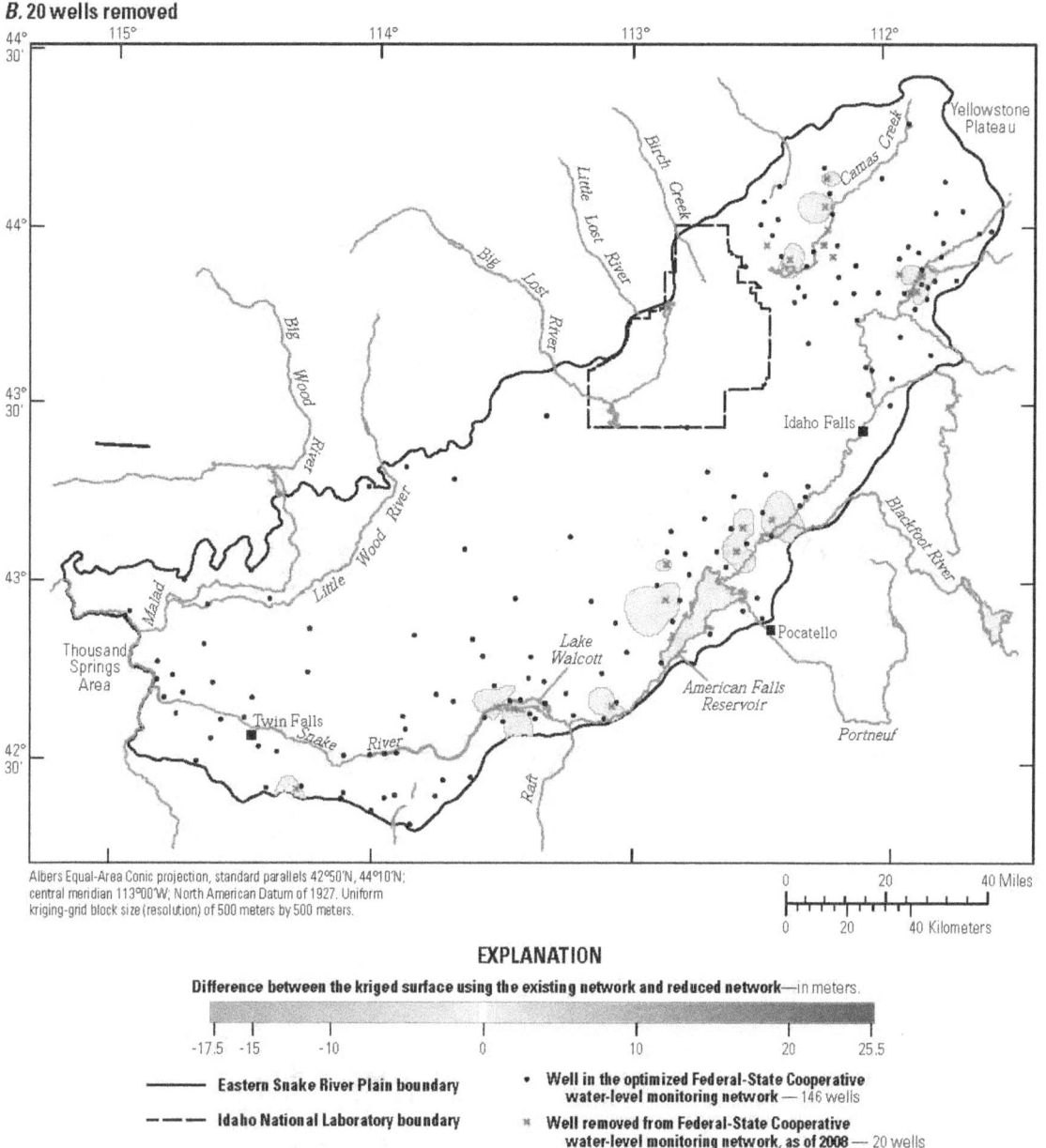

Albers Equal-Area Conic projection, standard parallels 42°50'N, 44°10'N;
central meridian 113°00'W; North American Datum of 1927. Uniform
kriging-grid block size (resolution) of 500 meters by 500 meters.

EXPLANATION

Difference between the kriged surface using the existing network and reduced network—in meters.

-17.5 -15   -10   0   10   20   25.5

——— Eastern Snake River Plain boundary
– – – Idaho National Laboratory boundary

• Well in the optimized Federal-State Cooperative
  water-level monitoring network — 146 wells

✳ Well removed from Federal-State Cooperative
  water-level monitoring network, as of 2008 — 20 wells

Figure 14.—Continued

C. 40 wells removed

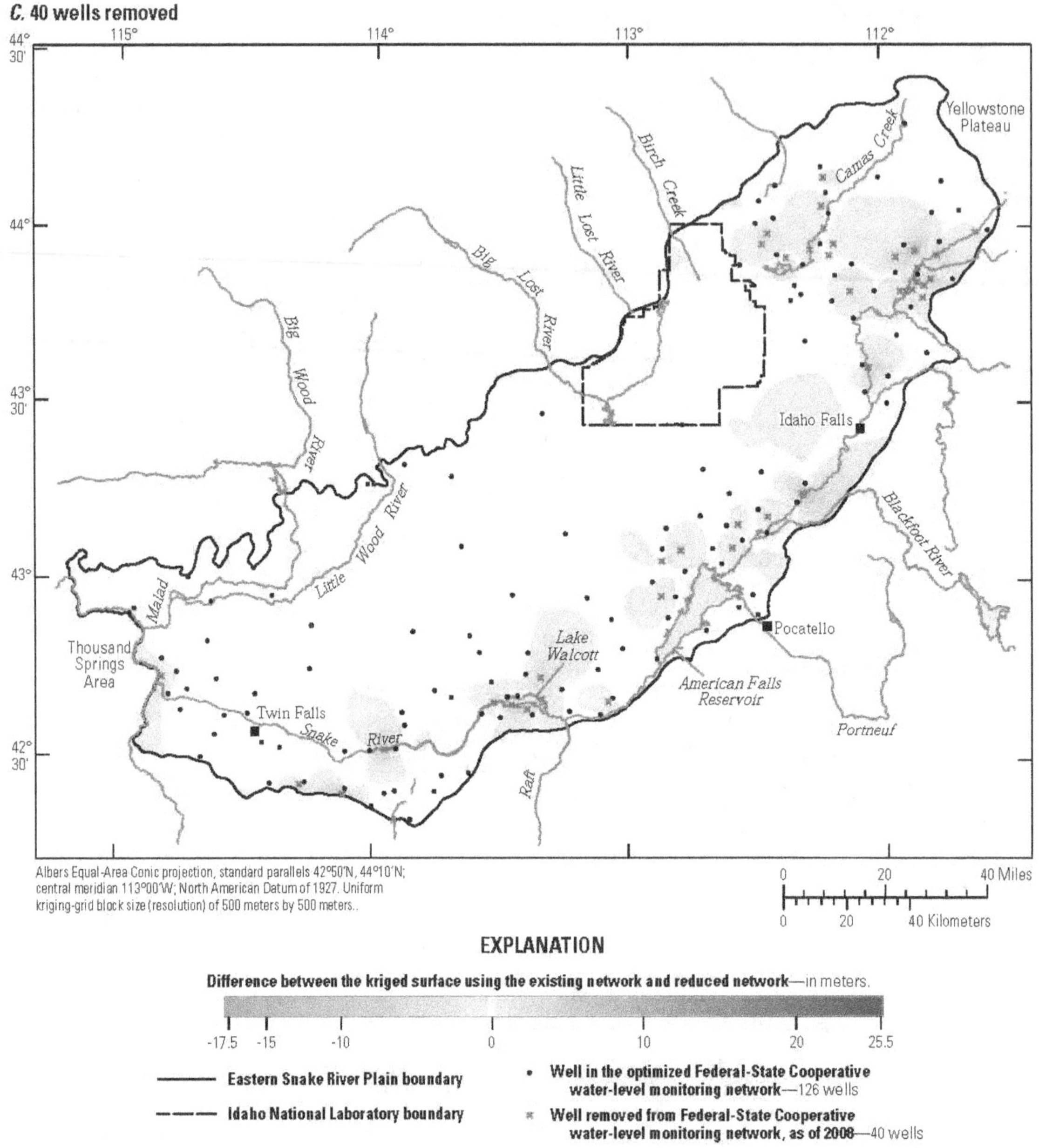
115°
114°
113°
112°
44° 30'
44°
43° 30'
43°
42° 30'
Yellowstone Plateau
Camas Creek
Birch Creek
Little Lost River
Big Lost River
Big Wood River
Little Wood River
Malad
Thousand Springs Area
Twin Falls
Snake River
Lake Walcott
Raft
American Falls Reservoir
Pocatello
Idaho Falls
Blackfoot River
Portneuf
Albers Equal-Area Conic projection, standard parallels 42°50'N, 44°10'N; central meridian 113°00'W; North American Datum of 1927. Uniform kriging-grid block size (resolution) of 500 meters by 500 meters.
0 20 40 Miles
0 20 40 Kilometers
EXPLANATION
Difference between the kriged surface using the existing network and reduced network—in meters.
-17.5 -15 -10 0 10 20 25.5
Eastern Snake River Plain boundary
Idaho National Laboratory boundary
Well in the optimized Federal-State Cooperative water-level monitoring network—126 wells
Well removed from Federal-State Cooperative water-level monitoring network, as of 2008—40 wells

Figure 14.—Continued

D. 60 wells removed

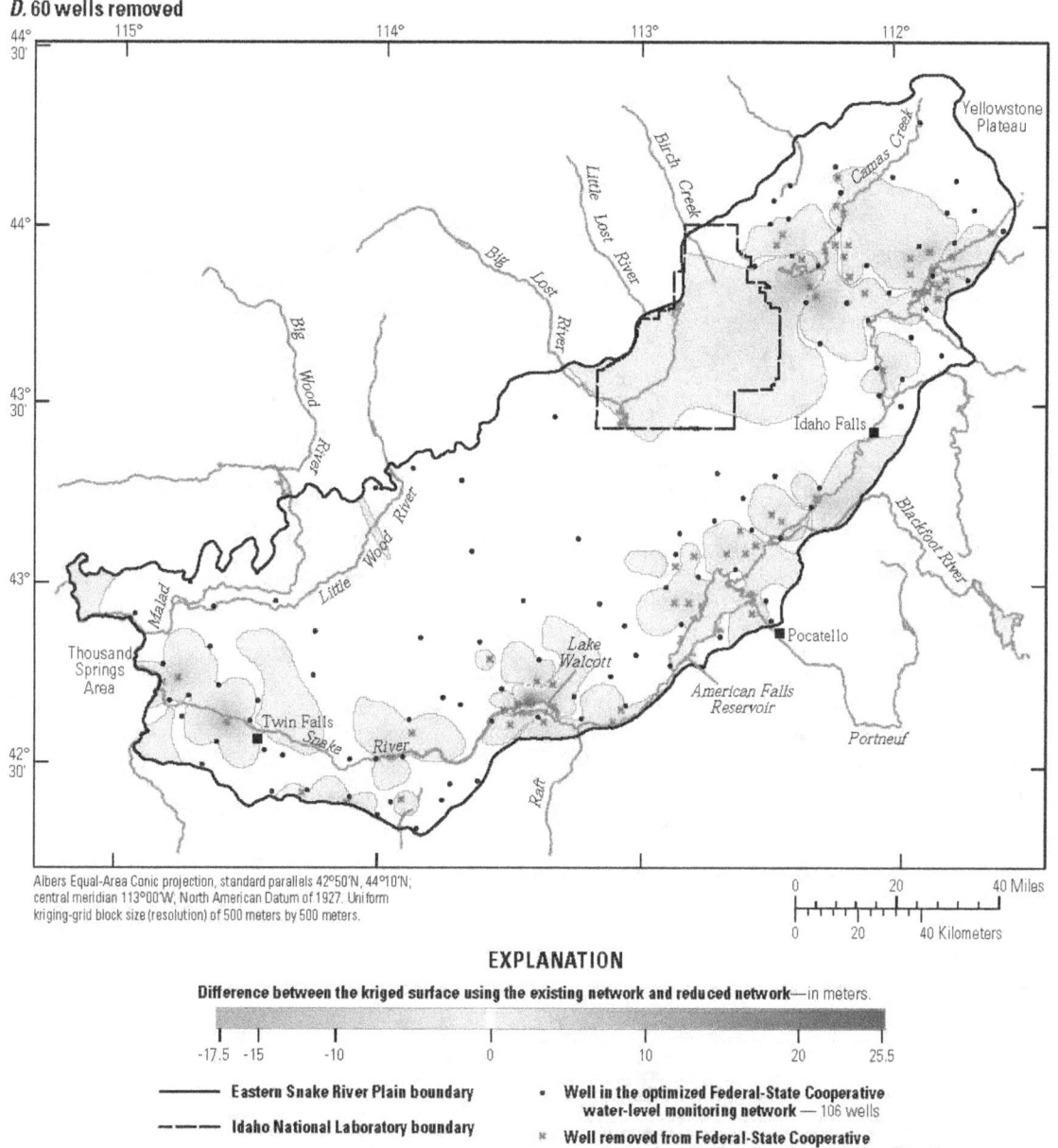

Albers Equal-Area Conic projection, standard parallels 42°50'N, 44°10'N;
central meridian 113°00'W, North American Datum of 1927. Uniform
kriging-grid block size (resolution) of 500 meters by 500 meters.

EXPLANATION

Difference between the kriged surface using the existing network and reduced network—in meters.

-17.5   -15   -10   0   10   20   25.5

——— Eastern Snake River Plain boundary

– – – Idaho National Laboratory boundary

• Well in the optimized Federal-State Cooperative
   water-level monitoring network — 106 wells

✕ Well removed from Federal-State Cooperative
   water-level monitoring network, as of 2008 — 60 wells

Figure 14.—Continued

E. **80 wells removed**

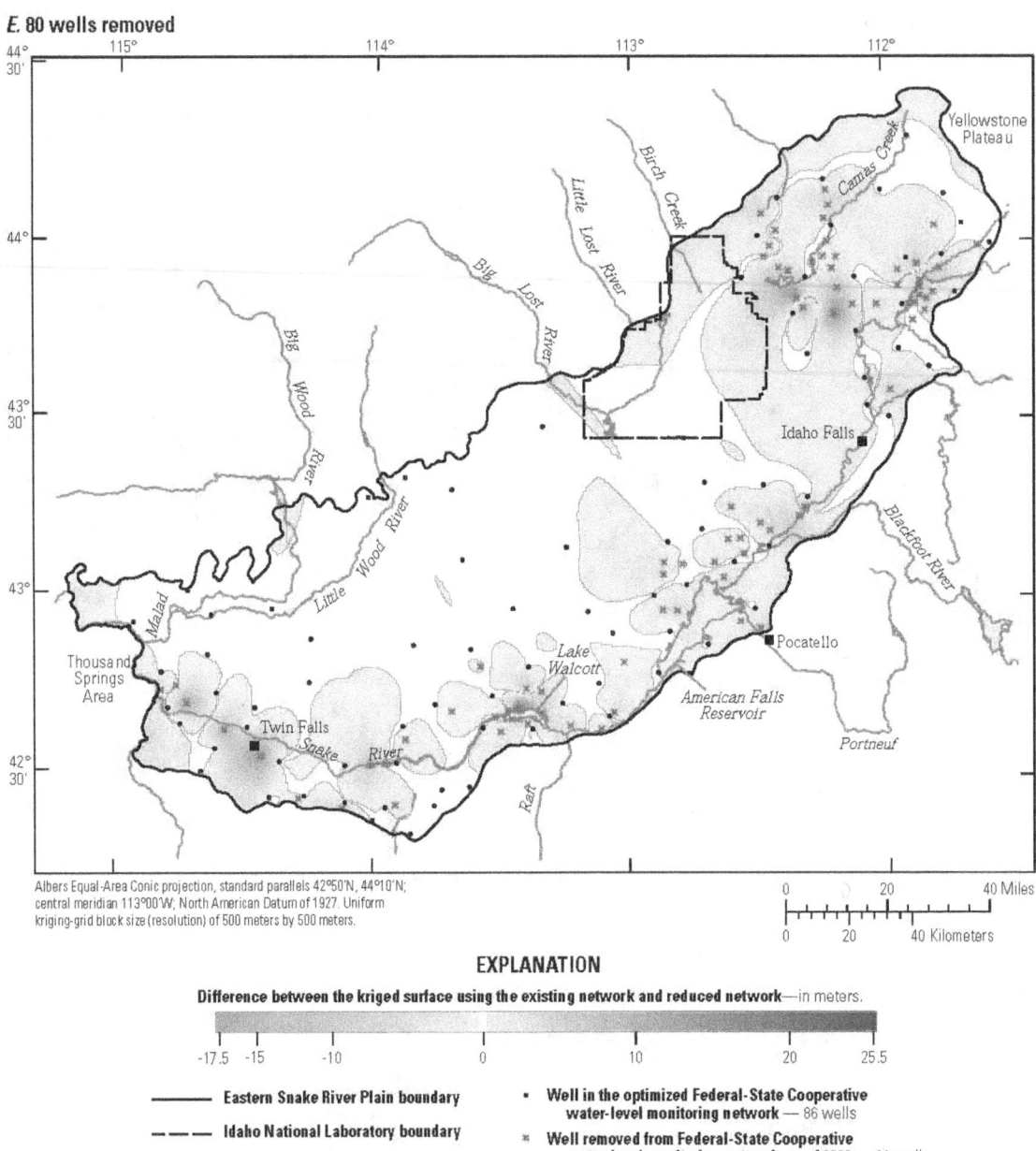

Albers Equal-Area Conic projection, standard parallels 42°50'N, 44°10'N; central meridian 113°00'W; North American Datum of 1927. Uniform kriging-grid block size (resolution) of 500 meters by 500 meters.

EXPLANATION

Difference between the kriged surface using the existing network and reduced network—in meters.

-17.5 -15   -10   0   10   20   25.5

——— Eastern Snake River Plain boundary

– – – Idaho National Laboratory boundary

• Well in the optimized Federal-State Cooperative water-level monitoring network — 86 wells

✳ Well removed from Federal-State Cooperative water-level monitoring network, as of 2008 — 80 wells

Figure 14.—Continued

A. 10 wells removed

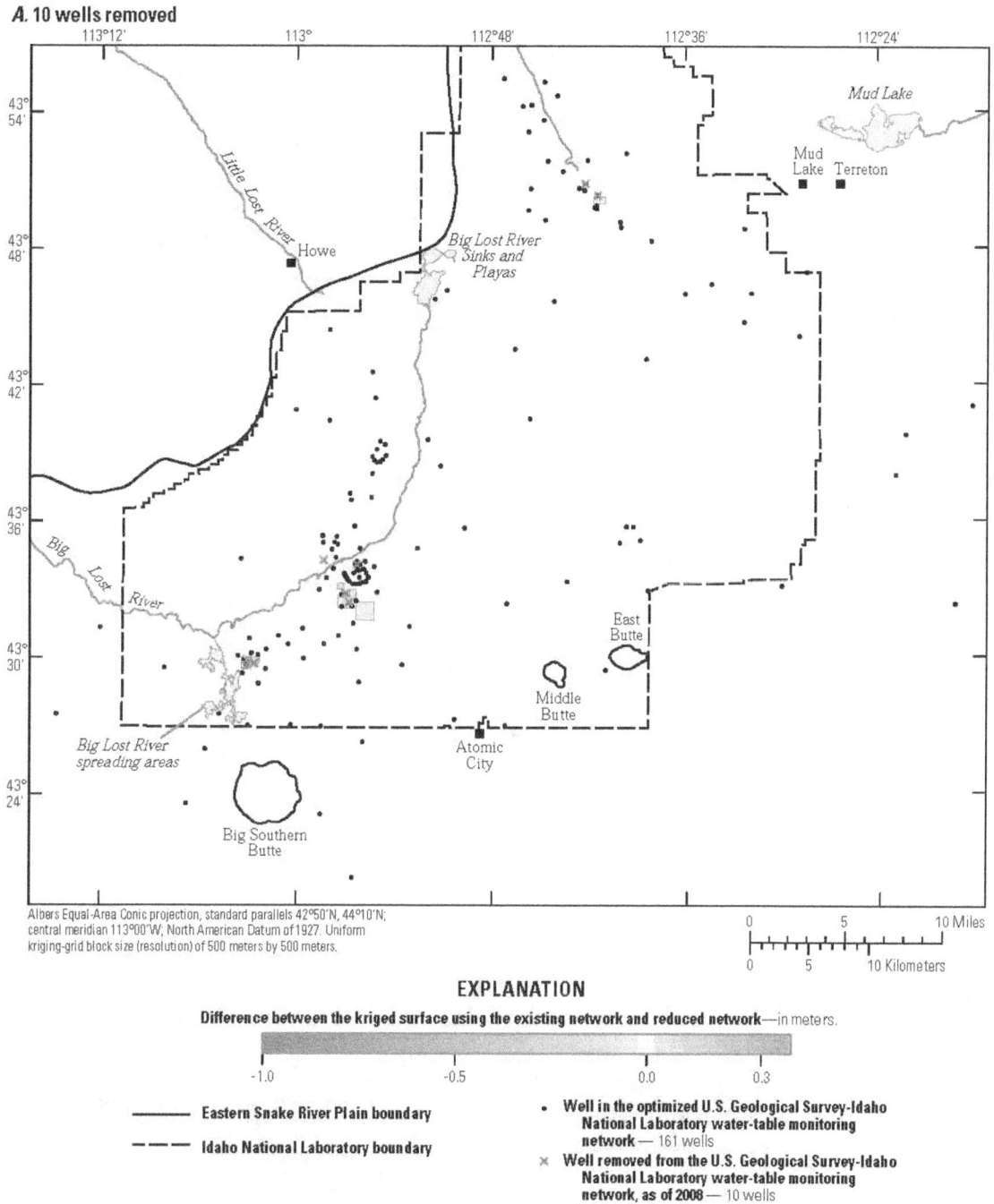

EXPLANATION

Difference between the kriged surface using the existing network and reduced network—in meters.

-1.0   -0.5   0.0   0.3

——— Eastern Snake River Plain boundary

‒ ‒ ‒ Idaho National Laboratory boundary

• Well in the optimized U.S. Geological Survey-Idaho National Laboratory water-table monitoring network — 161 wells

× Well removed from the U.S. Geological Survey-Idaho National Laboratory water-table monitoring network, as of 2008 — 10 wells

Figure 15. Difference between kriged water-table surfaces using the existing and reduced U.S. Geological Survey-Idaho National Laboratory water-level monitoring network, after removing (A) 10, (B) 20, (C) 40, (D) 60, and (E) 80 optimally selected wells, Idaho National Laboratory and vicinity, Idaho.

B. 20 wells removed

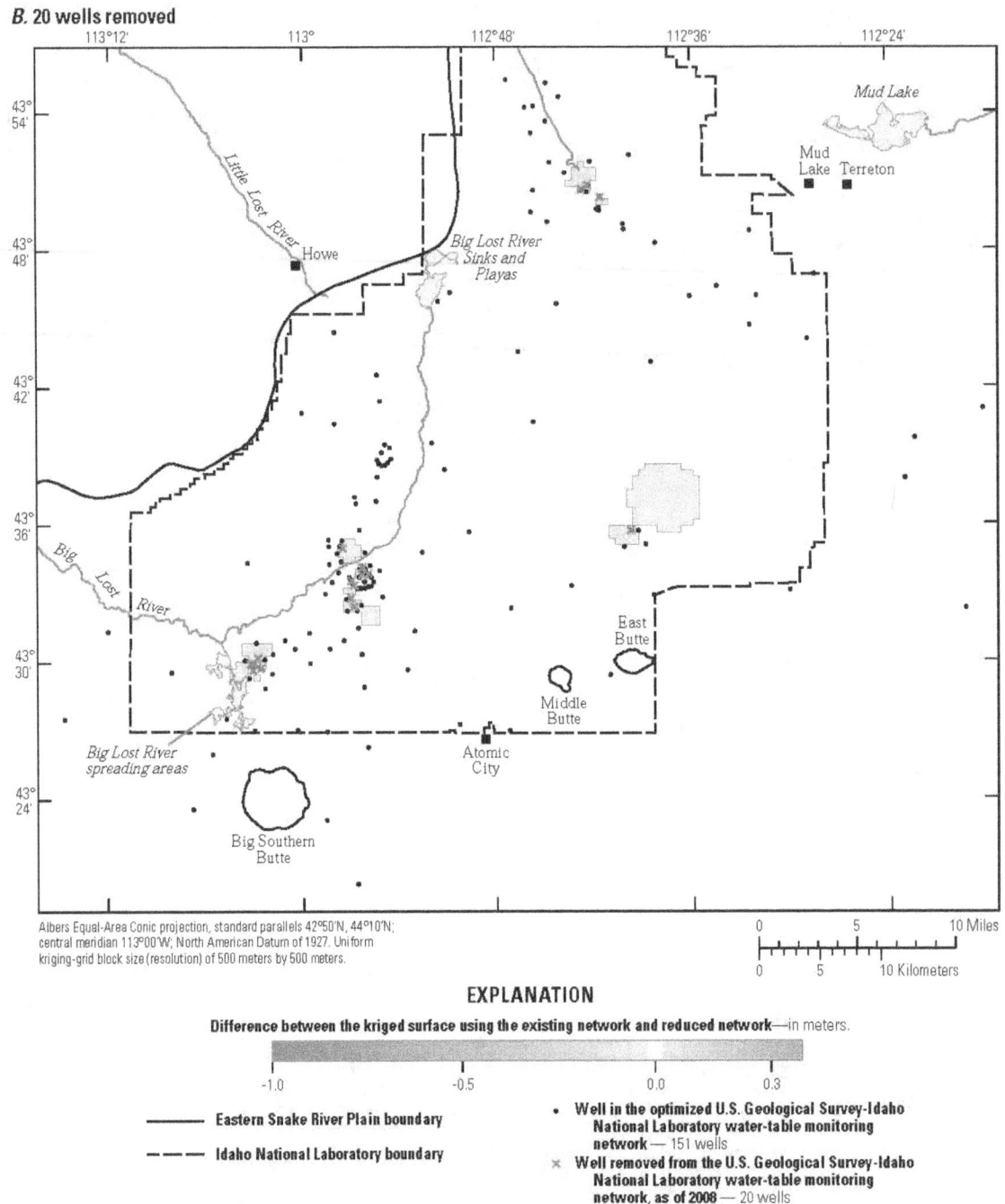

Albers Equal-Area Conic projection, standard parallels 42°50'N, 44°10'N;
central meridian 113°00'W; North American Datum of 1927. Uniform
kriging-grid block size (resolution) of 500 meters by 500 meters.

EXPLANATION

Difference between the kriged surface using the existing network and reduced network—in meters.

-1.0   -0.5   0.0   0.3

——— Eastern Snake River Plain boundary

- - - Idaho National Laboratory boundary

•   Well in the optimized U.S. Geological Survey-Idaho
    National Laboratory water-table monitoring
    network — 151 wells

×   Well removed from the U.S. Geological Survey-Idaho
    National Laboratory water-table monitoring
    network, as of 2008 — 20 wells

Figure 15.—Continued

C. 40 wells removed

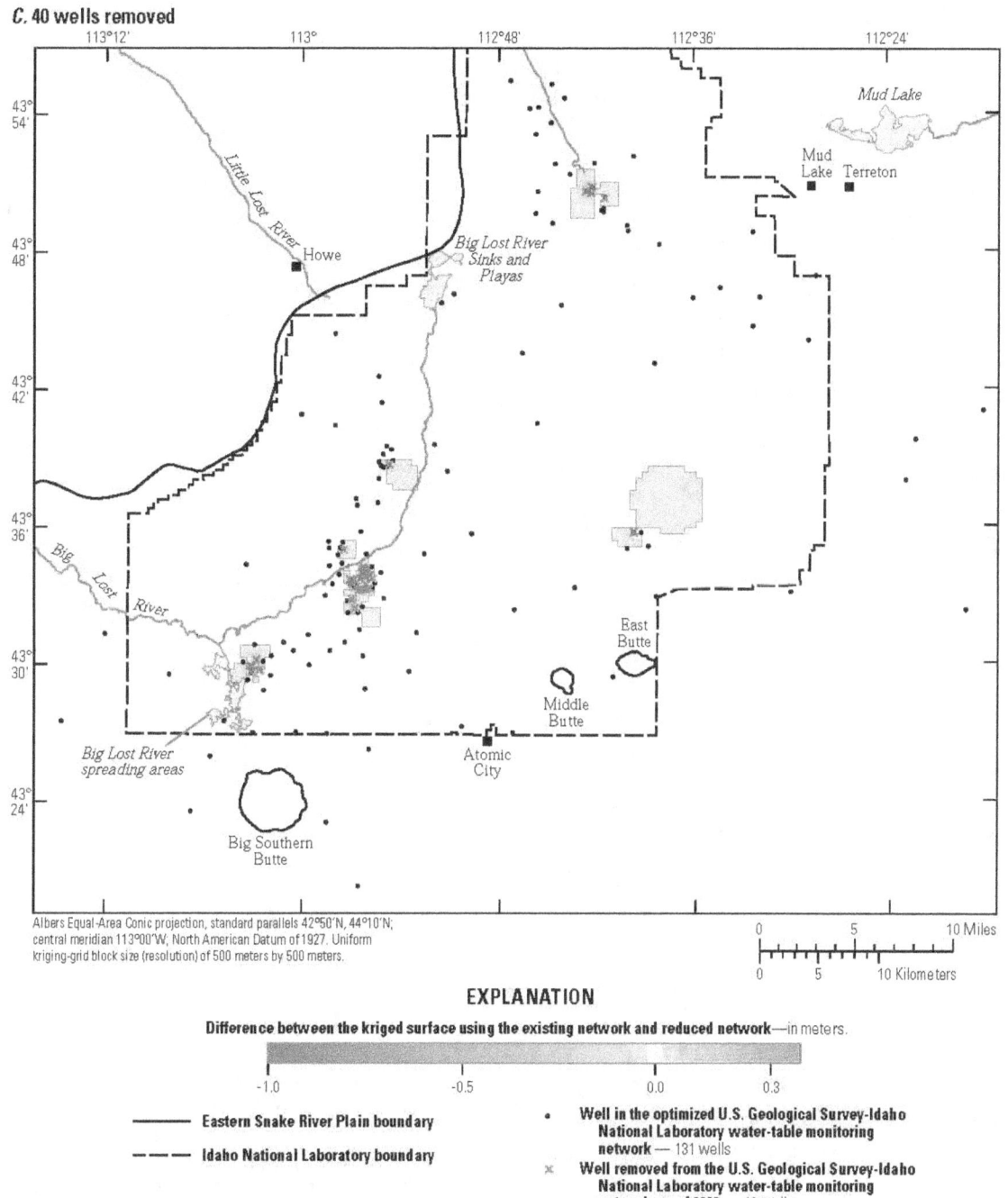

Albers Equal-Area Conic projection, standard parallels 42°50'N, 44°10'N;
central meridian 113°00'W; North American Datum of 1927. Uniform
kriging-grid block size (resolution) of 500 meters by 500 meters.

EXPLANATION

Difference between the kriged surface using the existing network and reduced network—in meters.

-1.0   -0.5   0.0   0.3

——— Eastern Snake River Plain boundary

- - - - Idaho National Laboratory boundary

• Well in the optimized U.S. Geological Survey-Idaho
National Laboratory water-table monitoring
network — 131 wells

× Well removed from the U.S. Geological Survey-Idaho
National Laboratory water-table monitoring
network, as of 2008 — 40 wells

Figure 15.—Continued

D. 60 wells removed

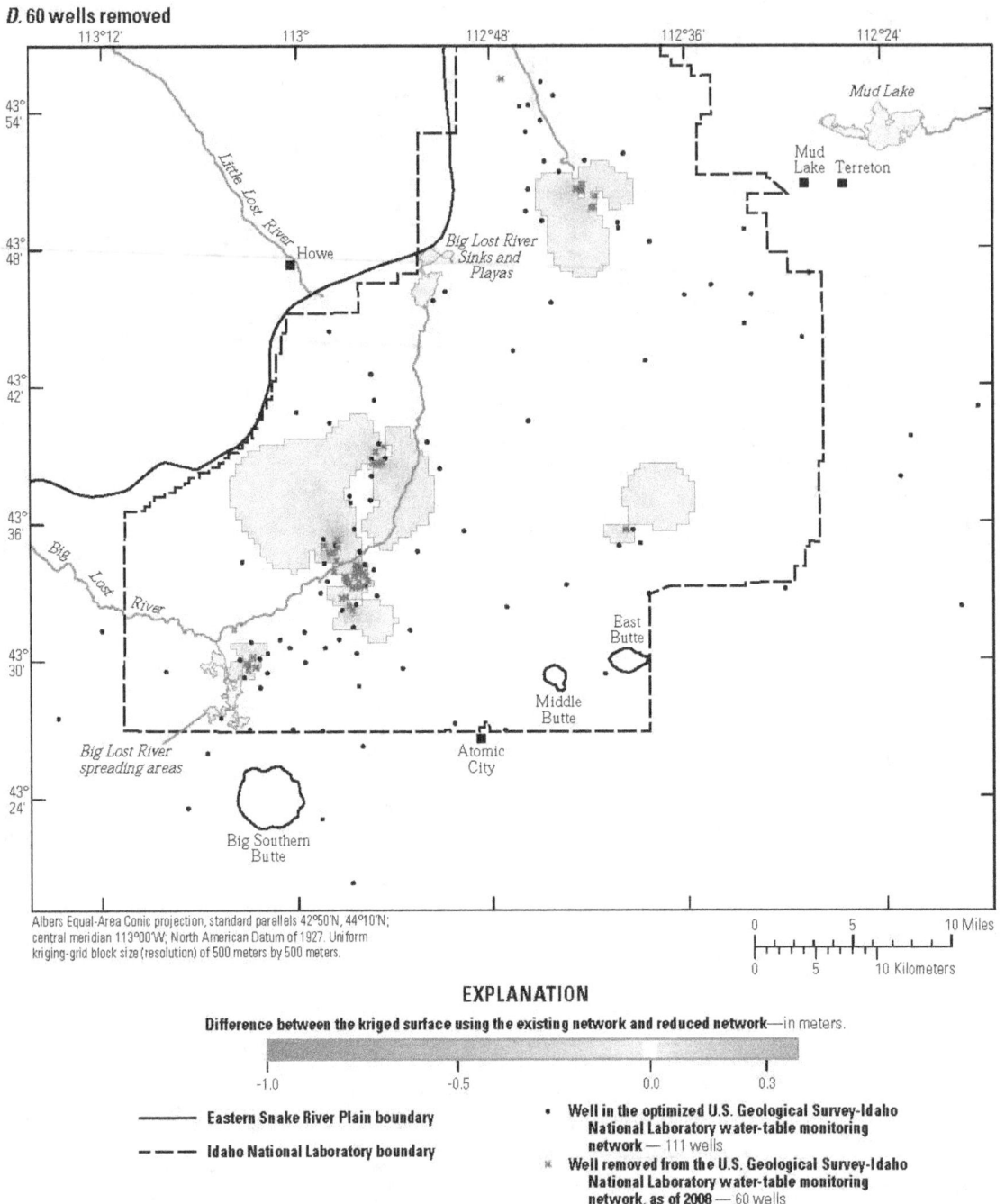

Albers Equal-Area Conic projection, standard parallels 42°50'N, 44°10'N; central meridian 113°00'W; North American Datum of 1927. Uniform kriging-grid block size (resolution) of 500 meters by 500 meters.

EXPLANATION

Difference between the kriged surface using the existing network and reduced network—in meters.

-1.0   -0.5   0.0   0.3

—— Eastern Snake River Plain boundary

– – – Idaho National Laboratory boundary

• Well in the optimized U.S. Geological Survey-Idaho National Laboratory water-table monitoring network — 111 wells

✳ Well removed from the U.S. Geological Survey-Idaho National Laboratory water-table monitoring network, as of 2008 — 60 wells

Figure 15.—Continued

E. 80 wells removed

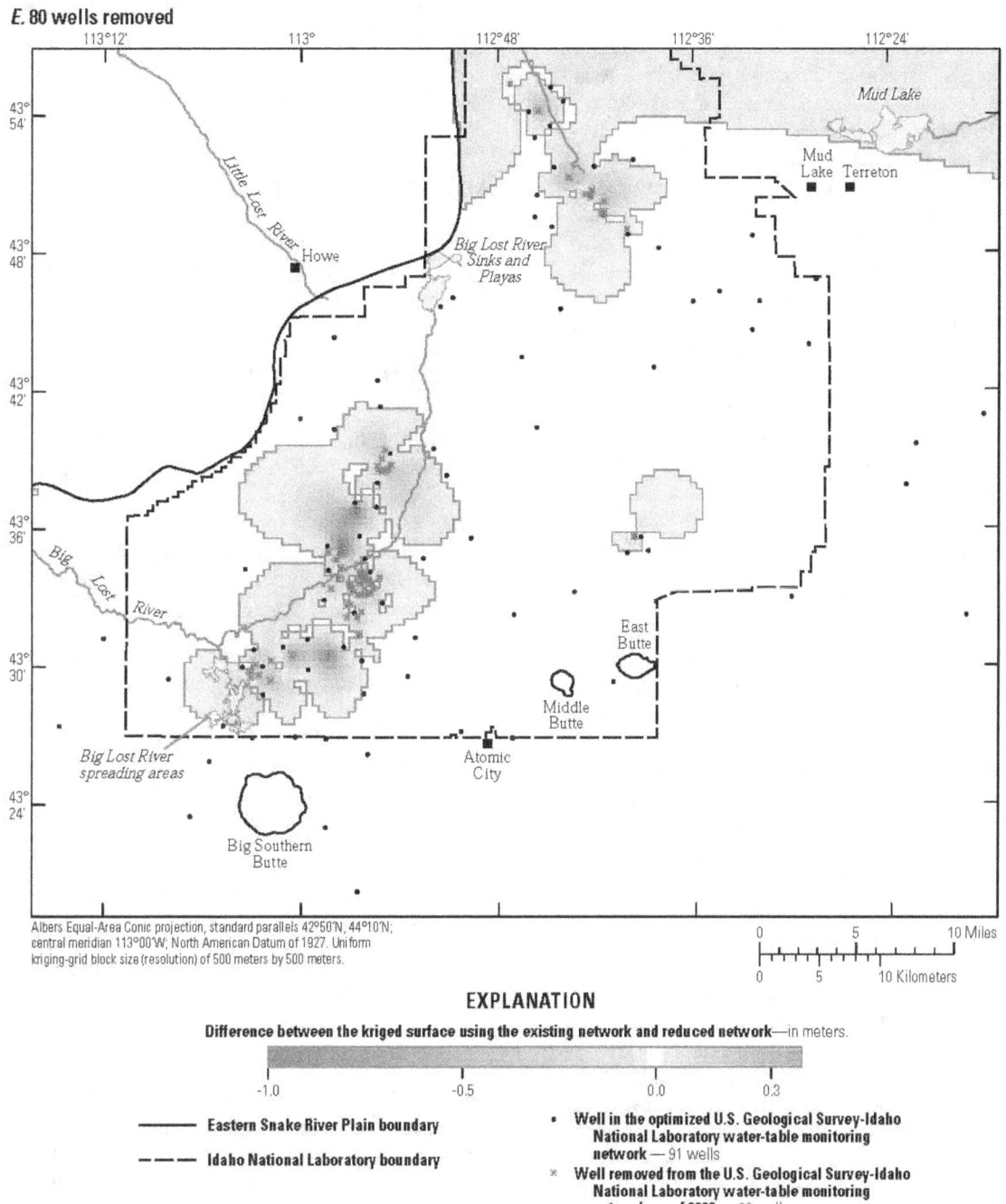

Albers Equal-Area Conic projection, standard parallels 42°50'N, 44°10'N;
central meridian 113°00'W; North American Datum of 1927. Uniform
kriging-grid block size (resolution) of 500 meters by 500 meters.

0   5   10 Miles

0   5   10 Kilometers

EXPLANATION

Difference between the kriged surface using the existing network and reduced network—in meters.

-1.0   -0.5   0.0   0.3

—— **Eastern Snake River Plain boundary**

‑ ‑ ‑ **Idaho National Laboratory boundary**

•   **Well in the optimized U.S. Geological Survey-Idaho National Laboratory water-table monitoring network** — 91 wells

×   **Well removed from the U.S. Geological Survey-Idaho National Laboratory water-table monitoring network, as of 2008** — 80 wells

Figure 15.—Continued

Summary and Conclusions

Budgetary constraints and the high cost of long-term groundwater level monitoring in the eastern Snake River Plain (ESRP) aquifer have necessitated a reduction in the number of observation wells in the existing networks. As of 2008, long-term groundwater water-level monitoring networks in the ESRP included a Federal-State Cooperative water-level monitoring network (Co-op network) with 166 observation wells, and a U.S. Geological Survey-Idaho National Laboratory water-level monitoring network (USGS-INL network) with 171 wells. The spatial distribution of observation wells in the Co-op network covers most of the ESRP, and USGS-INL network coverage is limited to the Idaho National Laboratory and vicinity. The planning objective for both networks is to reduce monitoring costs by removing observation wells that add little or no information characterizing the water table. To accomplish this objective, a reduced monitoring network was selected that satisfies the following design criteria: (1) interpolation error is minimized, (2) local anomalies in the water-table distribution are preserved, (3) variability of water-level measurements across time is preserved, and (4) measurement error is kept as small as possible. The total number of wells to remove from the existing network is left as a management decision.

The water-level monitoring networks were optimized using a genetic algorithm (GA) with universal kriging and statistical analysis. A series of GA runs were conducted for some of the control parameters to better understand the sensitivity of the algorithm to incremental changes in the parameters, and to determine reasonable settings for optimizing the existing monitoring networks. The network design tool is most sensitive to the number of wells removed from the original network and the spatial resolution of the kriging grid. As a compromise between solution accuracy and computational effort, existing water-level monitoring networks were optimized using the following control parameter settings: a kriging grid resolution of 2.5 kilometers for the Co-op network and 1.5 kilometers for the USGS-INL network, population size of 2,000, elitism rate of 0.05 (5 percent), crossover probability of 0.80 (80 percent), mutation probability of 0.30 (30 percent), and terminates after 50 consecutive iterations without any improvement in the best fitness. Each water-level monitoring network was optimized five times: by removing (1) 10, (2) 20, (3) 40, (4) 60, and (5) 80 observation wells from the original network. An examination of the trade-offs associated with changes in the number of wells to remove indicates that 20 wells (12 percent of the total number of wells in the original network) can be removed from the Co-op network with a relatively small degradation of the estimated water-table map, and 40 wells (23 percent) can be removed from the USGS-INL network before the water-table map degradation accelerates.

The optimal network designs indicate the robustness of the network design tool. Observation wells were removed from high well-density areas of the network while retaining the spatial pattern of the existing water-table map.

Acknowledgments

The author would like to thank R.J. Weakland of the USGS for her assistance during the early stages of this study.

References Cited

Ackerman, D.J., Rattray, G.W., Rousseau, J.P., Davis, L.C., and Orr, B.R., 2006, A conceptual model of ground-water flow in the eastern Snake River Plain aquifer at the Idaho National Laboratory and vicinity with implications for contaminant transport: U.S. Geological Survey Scientific Investigations Report 2006-5122 (DOE/ID-22198), 62 p. (Also available at http://pubs.usgs.gov/sir/2006/5122/.)

Anderson, S.R., and Liszewski, M.J., 1997, Stratigraphy of the unsaturated zone and the Snake River Plain aquifer at and near the Idaho National Engineering Laboratory, Idaho: U.S. Geological Survey Water-Resources Investigations Report 97-4183 (DOE/ID-22142), 65 p. (Also available at http://pubs.usgs.gov/wri/1997/4183/.)

Asefa, Tirusew, Kemblowski, M.W., Urroz, Gilberto, McKee, Mac, and Khalil, Abedalrazq, 2004, Support vectors-based groundwater head observation networks design: Water Resources Research, v. 40, W11509, 14 p.

Aziz, J.J., Ling, M., Rifai, H.S., Newell, C.J., and Gonzales, J.R., 2003, MAROS—A decision support system for optimizing monitoring plans: Ground Water, v. 41, no. 3, p. 355–367.

Bentley, P.J., and Wakefield, J.P., 1997, Finding acceptable solutions in the Pareto-optimal range using multiobjective genetic algorithms, in Chawdhry, P.K., and others, eds., part 5, Soft computing in engineering design and manufacturing: London, Springer-Verlag London Ltd., p. 231–240.

Bivand, R.S., Pebesma, E.J., and Gomez-Rubio, Virgilio, 2008, Applied spatial data analysis with R: New York, Springer, 374 p.

Bossong, C.R., Karlinger, M.R., Troutman, B.M., and Vecchia, A.V., 1999, Overview and technical and practical aspects for use of geostatistics in hazardous-, toxic-, and radioactive-waste-site investigations: U.S. Geological Survey Water-Resources Investigations Report 98-4145, 70 p. (Also available at http://pubs.usgs.gov/wri/1998/4145/.)

Cameron, Kirk, and Hunter, Philip, 2002, Using spatial models and kriging techniques to optimize long-term ground-water monitoring networks—A case study: Environmetrics, v. 13, p. 629–656.

David, Michel, 1976, The practice of kriging, *in* Guarascio, M., and others, eds., Advanced geostatistics in the mining industry: Boston, D. Reidell, p. 31–48.

Delfiner, Pierre, 1976, Linear estimation of nonstationary spatial phenomena, *in* Guarascio, M., and others, eds., Advanced geostatistics in the mining industry: Boston, D. Reidell, p. 49–68.

Dhar, Anirban, and Datta, Bithin, 2010, Logic-based design of groundwater monitoring network for redundancy reduction: Journal of Water Resources Planning and Management, v. 136, p. 88–94.

Fisher, J.C., and Twining, B.V., 2011, Multilevel groundwater monitoring of hydraulic head and temperature in the eastern Snake River Plain aquifer, Idaho National Laboratory, Idaho, 2007–08: U.S. Geological Survey Scientific Investigations Report 2010-5253, 62 p. (Also available at http://pubs.usgs.gov/sir/2010/5253/.)

Gangopadhyay, Subhrendu, Gupta, A.D., and Nachabe, M.H., 2001, Evaluation of ground water monitoring network by principal component analysis: Ground Water, v. 39, no. 2, p. 181–191.

Garabedian, S.P., 1992, Hydrology and digital simulation of the regional aquifer system, eastern Snake River Plain, Idaho: U.S. Geological Survey Professional Paper 1408-F, 102 p., 10 pls. (Also available at http://pubs.usgs.gov/pp/1408f/.)

Gardner, Martin, 1986, Knotted doughnuts and other mathematical entertainments—Chap. 2, The binary gray code: New York, W. H. Freeman, p. 11–27.

Grabow, G.L., Mote, C.R., Sanders, W.L., Smoot, J.L., and Yoder, D.C., 1993, Groundwater monitoring network design using minimum well density: Water Science and Technology, v. 28, no. 3–5, p. 327–335.

Haitjema, H.M., and Mitchell-Bruker, Sherry, 2005, Are water tables a subdued replica of the topography?: Ground Water, v. 43, no. 6, p. 781–786.

Herrera, G.S., and Pinder, G.F., 2005, Space-time optimization of groundwater quality sampling networks: Water Resources Research, v. 41, no. 12, 15 p.

Hijmans, R.J., and van Etten, Jacob, 2012, Geographic analysis and modeling with raster data—R package version 2.1-25: The Comprehensive R Archive Network Web site, accessed April 30, 2013, at http://CRAN.R-project.org/package=raster.

Holland, J.H., 1975, Adaptation in natural and artificial systems: Ann Arbor, Mich., University of Michigan Press, 183 p.

Isaaks, E.H., and Srivastava, R.M., 1989, An introduction to applied geostatistics: New York, Oxford University Press, 561 p.

Keitt, T.H., Bivand, Roger, Pebesma, Edzer, and Rowlingson, Barry, 2012, rgdal—Bindings for the geospatial data abstraction library—R package version 0.8-8: The Comprehensive R Archive Network Web site, accessed April 30, 2013, at http://CRAN.R-project.org/package=rgdal.

Khan, Shahbaz, Chen, H.F., and Rana, Tariq, 2008, Optimizing ground water observation networks in irrigation areas using principal component analysis: Ground Water Monitoring and Remediation, v. 28, no. 3, p. 93–100.

Kitanidis, P.K., 1997, Introduction to geostatistics—Applications to hydrogeology: New York, Cambridge University Press, 249 p.

Koza, J.R., 1992, Genetic programming—Vol. 1, on the programming of computers by means of natural selection (complex adaptive systems): London, A Bradford Book, 819 p.

Li, Yuanhai, and Hilton, A.B.C, 2007, Optimal groundwater monitoring design using an ant colony optimization paradigm: Environmental Modeling and Software, v. 22, p. 110–116.

Lin, Yu-Pin, and Rouhani, Shaharokh, 2001, Multiple-point variance analysis for optimal adjustment of a monitoring network: Environmental Monitoring and Assessment, v. 69, p. 239–266.

Lindholm, G.F., Garabedian, S.P., Newton, G.D., and Whitehead, R.L., 1988, Configuration of the water table and depth to water, spring 1980, water-level fluctuations, and water movement in the Snake River Plain regional aquifer system, Idaho and eastern Oregon: U.S. Geological Survey Hydrologic Atlas HA-703, scale 1:500,000.

Loaiciga, H.A., Charbeneau, R.J., Everett, L.G., Fogg, G.E., Hobbs, B.F., and Rouhani, Shahrokh, 1992, Review of ground-water quality monitoring network design: Journal of Hydraulic Engineering, v. 188, no. 1, p. 11–37.

Marczyk, Adam, 2004, Genetic algorithms and evolutionary computation: The TalkOrigins Archive Web site, accessed April 1, 2013, at http://www.talkorigins.org/faqs/genalg/genalg.html.

Mundorff, M.J., Crosthwaite, E.G., and Kilburn, Chabot, 1964, Ground water for irrigation in the Snake River Basin in Idaho: U.S. Geological Survey Water-Supply Paper 1654, 224 p.

Nunes, L.M., Cunha, M.C., and Ribeiro, L., 2004a, Groundwater monitoring network optimization with redundancy reduction: Journal of Water Resources Planning and Management, v. 130, p. 33–43.

Nunes, L.M., Paralta, E., Cunha, M.C., and Ribeiro, Luís, 2004b, Groundwater nitrate monitoring network optimization with missing data: Water Resources Research, v. 40, W02406, 18 p.

Passarella, Giuseppe, Vurro, Michele, Agostino, V.D., and Barcelona, M.J., 2003, Cokriging optimization of monitoring network configuration based on fuzzy and non-fuzzy variogram evaluation: Environmental Monitoring and Assessment, v. 82, p. 1–21.

Pebesma, E.J., 2004, Multivariable geostatistics in S— The gstat package: Computers and Geosciences, v. 30, p. 683–691.

Pebesma, E.J., and Bivand, R.S., 2005, Classes and methods for spatial data in R: R News, v. 5, no. 2, p. 9–13, accessed April 1, 2013, at http://cran.r-project.org/doc/Rnews/.

Philip, R.D., and Kitanidis, P.K., 1989, Geostatistical estimation of hydraulic head gradients: Ground Water, v. 27, no. 6, p. 855–865.

Pierce, K.L., and Morgan, L.A., 1992, The track of the Yellowstone hotspot: Volcanism, uplift and faulting, in Link, P.K., Kuntz, M.A., and Platt, L.B., eds., Regional geology of Eastern Idaho and Western Wyoming: Geological Society of America Memoir 179, p. 1–53.

R Development Core Team, 2013, R—A language and environment for statistical computing: R Foundation for Statistical Computing, Vienna, Austria, accessed April 1, 2013, at http://www.R-project.org.

Reed, Patrick, Minsker, Barbara, and Valocchi, A.J., 2000, Cost-effective long-term groundwater monitoring design using a genetic algorithm and global mass interpolation: Water Resources Research, v. 36, no. 12, p. 3731–3741.

Rodgers, D.W., Ore, H.T., Bobo, R.T., McQuarrie, Nadine, and Zentner, Nick, 2002, Extension and subsidence of the eastern Snake River Plain, Idaho, in Bonnichsen, Bill, White, C.M., and McCurry, Michael, eds., Tectonic and magmatic evolution of the Snake River Plain volcanic province: Idaho Geological Survey Bulletin 30, p. 121–155.

Scrucca, Luca, 2013, GA—A package for genetic algorithms in R: Journal of Statistical Software, v. 53, no. 4, 37 p.

Snyder, D.T., 2008, Estimated depth to ground water and configuration of the water table in the Portland, Oregon area: U.S. Geological Survey Scientific Investigations Report 2008-5059, 40 p. (Also available at http://pubs.usgs.gov/sir/2008/5059/.)

Stearns, H.T., Crandall, Lynn, and Steward, W.G., 1938, Geology and ground-water resources of the Snake River Plain in southeastern Idaho: U.S. Geological Survey Water-Supply Paper 774, 268 p.

Theodossiou, Nicolaos, and Latinopoulos, Pericles, 2006, Evaluation and optimisation of groundwater observation networks using the Kriging methodology: Environmental Modeling and Software, v. 21, no. 7, p. 991–1000.

Whitehead, R.L., 1992, Geohydrologic framework of the Snake River Plain regional aquifer system, Idaho and eastern Oregon: U.S. Geological Survey Professional Paper 1408-B, 32 p., 6 pls. (Also available at http://pubs.usgs.gov/pp/1408b/.)

Yeh, M.-S., Lin, Y.-P, and Chang, L.-C., 2006, Designing an optimal multivariate geostatistical groundwater quality monitoring network using factorial kriging and genetic algorithms: Environmental Geology, v. 50, p. 101–121.

Table 5   53

Table 5. Wells in the Federal-State Cooperative and U.S. Geological Survey-Idaho National Laboratory water-level monitoring networks, eastern Snake River Plain, Idaho, during 2008.

[**Local name:** local well identifier used in this study. **Map No.:** identifier used to locate wells on maps located in figures and as a cross reference with data in other tables. **Site No.:** unique numerical identifiers used to access well data (http://waterdata.usgs.gov/nwis). **Network name:** name(s) of long-term water-table monitoring network, as of 2008. **Longitude and Latitude:** in degrees, minutes, and seconds, and are in conformance with the North American Datum of 1983. **Reference point elevation:** land-surface reference point, in meters above North American Vertical Datum of 1988 (NAVD 88). **Reference point location error:** accuracy of reference point. **Sample size:** number of water-level measurements recorded for the period-of-record (POR) and calendar year 2008. **2008 water-level elevation:** median water-level elevation for calendar year 2008 (z), in meters above the NAVD 88. **2008 measurement method error:** accuracy of method used to measure depth to water during calendar year 2008. **2008 measurement error:** mean accuracy of water-level elevation measurements for calendar year 2008 (ε_z). **Standard deviation:** of water-level elevation measurements for the entire period-of-record (σ_z); duration varies for each well site. **Estimation error:** determined from leave-one-out cross validation ($z - z^*$). **Abbreviations:** Co-op, Federal-State Cooperative water-level monitoring network; USGS-INL, U.S. Geological Survey-Idaho National Laboratory water-level monitoring network; m, meter]

Local name	Map No.	Site No.	Network name	Longitude	Latitude	Reference point elevation (m)	Reference point location error (m)	Sample size POR	Sample size 2008	2008 water-level elevation (m)	2008 measurement method error (m)	2008 measurement error (m)	Standard deviation (m)	Estimation error (m)
12S 22E 35BCC1	1	422013113510501	Co-op	113°51'07"	42°20'12"	1,338.18	1.524	196	6	1,187.09	0.003	1.53	7.92	-33.17
12S 21E 16DCC1	2	422227113595901	Co-op	114°00'02"	42°22'26"	1,335.41	0.030	322	6	1,293.11	0.003	0.03	3.87	90.91
12S 20E 04DBC1	3	422424114070001	Co-op	114°07'03"	42°24'23"	1,317.73	0.152	507	10	1,240.17	0.003	0.16	3.32	5.67
12S 21E 02DAA1	4	422434113570201	Co-op	113°57'04"	42°24'34"	1,330.30	0.030	1,830	6	1,184.20	0.003	0.03	8.75	-33.91
11S 23E 34CDC1	5	422458113452701	Co-op	113°45'30"	42°24'57"	1,302.83	0.030	316	6	1,190.30	0.003	0.03	5.54	-50.10
11S 22E 32CCC1	6	422501113543901	Co-op	113°54'42"	42°25'00"	1,314.59	0.030	400	6	1,183.58	0.003	0.03	5.32	-0.85
11S 20E 33DAD1	7	422518114062701	Co-op	114°06'30"	42°25'19"	1,294.87	0.003	694	2	1,221.60	0.003	0.01	10.31	-10.09
11S 18E 25DDC1	8	422555114172101	Co-op	114°17'21"	42°25'55"	1,263.45	0.762	8	2	1,238.56	0.003	0.77	0.36	0.75
11S 17E 25DDD2	9	422600114240901	Co-op	114°24'12"	42°25'59"	1,262.38	0.003	853	4	1,234.74	0.003	0.01	1.42	24.09
11S 19E 30ADD1	10	422621114160501	Co-op	114°16'08"	42°26'20"	1,267.96	0.003	576	6	1,236.94	0.003	0.01	1.62	5.93
11S 23E 14DDD1	11	422739113434001	Co-op	113°43'43"	42°27'38"	1,290.29	1.524	180	6	1,279.36	0.003	1.53	1.23	74.59
11S 24E 14BDB1	12	422810113372001	Co-op	113°37'23"	42°28'09"	1,329.93	3.048	38	2	1,233.04	0.003	3.05	9.72	-29.08
11S 15E 02BBB1	13	423018114401701	Co-op	114°40'20"	42°30'17"	1,263.42	1.524	234	4	1,195.03	0.003	1.53	0.91	51.70
10S 20E 27BCC1	14	423134114062601	Co-op	114°06'31"	42°31'34"	1,275.68	0.003	203	2	1,165.95	0.003	0.01	3.70	-11.17
10S 21E 28BCB1	15	423145114003001	Co-op	114°00'33"	42°31'44"	1,268.09	0.003	409	10	1,165.71	0.003	0.01	3.53	-10.71
10S 21E 26AAA2	16	423159113570302	Co-op	113°57'06"	42°31'58"	1,268.62	0.762	274	9	1,181.32	0.003	0.77	2.48	4.89
10S 22E 20CDC1	17	423206113542301	Co-op	113°54'26"	42°32'05"	1,265.73	0.003	408	9	1,186.31	0.003	0.01	3.87	-6.19
10S 18E 20DDD1	18	423207114215301	Co-op	114°21'56"	42°32'06"	1,195.43	1.524	376	2	1,141.39	0.003	1.53	1.20	-20.47
10S 17E 14CCD1	19	423255114260601	Co-op	114°26'09"	42°32'54"	1,155.48	1.524	226	4	1,135.72	0.003	1.53	4.21	14.10
10S 16E 07DAC1	20	423406114370301	Co-op	114°37'07"	42°34'06"	1,153.04	1.524	220	4	1,143.32	0.003	1.53	1.32	16.76
09S 22E 33ADA1	21	423604113522401	Co-op	113°52'27"	42°36'03"	1,278.80	0.003	211	2	1,202.09	0.003	0.01	2.43	1.92
09S 16E 21DCD1	22	423722114345101	Co-op	114°34'54"	42°37'21"	1,081.39	1.524	204	6	1,076.95	0.003	1.53	0.50	11.59
09S 25E 23DBA1	23	423732113295801	Co-op	113°30'01"	42°37'31"	1,301.54	0.003	2,448	11	1,253.88	0.003	0.01	2.60	4.24
09S 17E 20CAA1	24	423747114293101	Co-op	114°29'32"	42°37'46"	1,107.91	1.524	218	6	1,013.45	0.003	1.53	1.90	-80.54
09S 26E 13CCC2	25	423802113222701	Co-op	113°22'26"	42°38'01"	1,305.50	0.762	9	2	1,253.51	0.003	0.77	0.91	-4.34
09S 29E 18CDA1	26	423808113063601	Co-op	113°06'38"	42°38'07"	1,296.18	0.003	128	3	1,280.78	0.003	0.01	1.60	-2.88
09S 25E 18DDA1	27	423811113341201	Co-op	113°34'15"	42°38'10"	1,267.71	0.762	434	11	1,253.27	0.003	0.77	2.57	6.07
09S 14E 13DDD1	28	423814114450901	Co-op	114°45'12"	42°38'13"	1,071.94	3.048	216	4	1,062.50	0.003	3.05	7.98	46.35
09S 22E 16CDB1	29	423817113530201	Co-op	113°53'05"	42°38'16"	1,281.41	1.524	172	2	1,198.66	0.003	1.53	3.17	-2.52
09S 28E 18B AD1	30	423837113134301	Co-op	113°13'44"	42°38'36"	1,286.25	0.003	215	3	1,281.20	0.003	0.01	0.44	3.14
09S 26E 10DDD1	31	423855113233901	Co-op	113°23'42"	42°38'54"	1,286.35	0.030	1,022	6	1,253.89	0.003	0.03	2.91	-5.86

Table 5. Wells in the Federal-State Cooperative and U.S. Geological Survey-Idaho National Laboratory water-level monitoring networks, eastern Snake River Plain, Idaho, during 2008.—Continued

Local name	Map No.	Site No.	Network name	Longitude	Latitude	Reference point elevation (m)	Reference point location error (m)	Sample size POR	Sample size 2008	2008 water-level elevation (m)	2008 measurement method error (m)	2008 measurement error (m)	Standard deviation (m)	Estimation error (m)
09S 26E 07AAB1	32	423943113272001	Co-op	113°27'23"	42°39'42"	1,281.10	0.152	264	6	1,252.98	0.003	0.16	3.08	-3.84
09S 25E 03CAC1	33	424003113313101	Co-op	113°31'34"	42°40'02"	1,267.92	0.003	741	9	1,248.31	0.003	0.01	1.96	1.18
09S 29E 04BCA1	34	424013113043801	Co-op	113°04'44"	42°40'13"	1,289.27	0.003	770	3	1,287.65	0.003	0.01	0.84	-1.49
08S 27E 31DDA1	35	424042113201101	Co-op	113°20'13"	42°40'39"	1,281.87	0.003	921	6	1,270.67	0.003	0.01	1.38	1.85
08S 29E 34CBC1	36	424052113033901	Co-op	113°03'41"	42°40'51"	1,338.87	0.003	230	3	1,292.34	0.003	0.01	0.98	1.65
08S 24E 31DAC1	37	424053113412801	Co-op	113°41'31"	42°40'52"	1,289.20	0.003	2,642	10	1,235.85	0.003	0.01	2.29	-1.41
09S 14E 03BAA1	38	424053114480301	Co-op	114°48'07"	42°40'52"	978.95	0.030	339	6	955.57	0.003	0.03	0.59	-40.69
08S 25E 36DAA1	39	424102113282101	Co-op	113°28'24"	42°41'01"	1,283.86	0.003	1,198	9	1,244.72	0.003	0.01	2.75	-12.64
08S 17E 33DAD2	40	424105114274901	Co-op	114°27'50"	42°41'06"	1,165.84	1.524	190	6	1,086.23	0.003	1.53	1.07	49.75
08S 26E 33BCB2	41	424112113255402	Co-op	113°25'57"	42°41'11"	1,284.99	0.003	607	6	1,277.74	0.003	0.01	0.48	28.26
08S 15E 32BBA1	42	424144114434101	Co-op	114°43'43"	42°41'44"	1,010.96	1.524	8	2	984.82	0.003	1.53	1.89	-18.92
08S 23E 27BDC1	43	424201113452701	Co-op	113°45'30"	42°42'00"	1,291.63	0.003	196	2	1,229.13	0.003	0.01	2.86	4.82
08S 27E 23DDD1	44	424221113152501	Co-op	113°15'28"	42°42'22"	1,310.52	0.003	130	5	1,285.86	0.003	0.01	0.96	6.92
08S 16E 17CCC1	45	424331114365001	Co-op	114°36'54"	42°43'30"	1,064.61	1.524	132	7	1,012.17	0.003	1.53	1.74	-23.53
08S 25E 16DAC1	46	424334113320201	Co-op	113°32'05"	42°43'33"	1,294.36	0.003	132	2	1,238.59	0.003	0.01	3.29	-3.68
08S 14E 16CBB1	47	424353114494701	Co-op	114°49'49"	42°43'52"	968.66	0.003	2,722	6	955.92	0.003	0.01	0.35	2.14
08S 27E 07DBC1	48	424419113201801	Co-op	113°20'25"	42°44'19"	1,319.25	3.048	65	5	1,265.42	0.003	3.05	0.78	4.01
08S 14E 12CBC1	49	424439114461201	Co-op	114°46'16"	42°44'39"	998.15	1.524	206	6	975.97	0.003	1.53	1.00	5.84
08S 26E 03DCC1	50	424454113240101	Co-op	113°24'04"	42°44'53"	1,325.81	0.003	623	9	1,240.68	0.003	0.01	2.34	-16.43
08S 19E 05DAB1	51	424529114150901	Co-op	114°15'12"	42°45'28"	1,243.15	0.003	999	6	1,152.31	0.003	0.01	1.89	17.71
08S 28E 01AAA2	52	424543113071002	Co-op	113°07'13"	42°45'43"	1,371.09	0.152	114	3	1,299.60	0.003	0.16	1.12	3.60
07S 14E 33BBB1	53	424653114494601	Co-op	114°49'49"	42°46'51"	997.84	1.524	190	6	964.61	0.003	1.53	1.07	10.39
07S 30E 24DDC1	54	424730112531701	Co-op	112°53'20"	42°47'29"	1,340.41	0.003	306	2	1,322.23	0.003	0.01	1.94	6.41
07S 26E 14CCC1	55	424826113233201	Co-op	113°23'35"	42°48'25"	1,343.06	0.003	677	8	1,241.24	0.003	0.01	2.26	-5.83
07S 25E 19BAA1	56	424828113345201	Co-op	113°34'55"	42°48'27"	1,317.83	0.003	2,289	10	1,236.17	0.003	0.01	2.25	0.88
07S 29E 12CCC2	57	424916113012001	Co-op	113°01'20"	42°49'16"	1,392.45	1.524	7	3	1,312.98	0.003	1.53	0.92	0.51
07S 15E 12CBA3	58	424955114390303	Co-op	114°39'06"	42°49'54"	1,098.10	0.003	137	6	1,043.17	0.003	0.01	1.97	22.76
06S 24E 32DBA1	59	425118113370801	Co-op	113°37'12"	42°51'18"	1,321.05	0.305	30	2	1,235.00	0.003	0.31	1.41	1.65
06S 22E 28CDD1	60	425155113503901	Co-op	113°50'42"	42°51'54"	1,287.99	0.003	120	5	1,220.44	0.003	0.01	2.57	-3.32
06S 32E 27ADC1	61	425216112414301	Co-op	112°41'45"	42°52'15"	1,347.25	0.003	2,045	6	1,336.33	0.008	0.01	0.80	1.41
06S 19E 19CCD1	62	425250114145101	Co-op	114°14'54"	42°52'49"	1,232.28	3.048	35	1	1,158.72	0.003	3.05	1.96	-9.99
06S 29E 15BBC1	63	425412113035601	Co-op	113°04'00"	42°54'09"	1,442.76	3.048	174	3	1,314.94	0.003	3.05	1.03	6.67
06S 13E 18ABC1	64	425421114572901	Co-op	114°57'32"	42°54'20"	863.39	3.658	526	6	866.33	0.003	3.66	7.61	-88.53
06S 31E 16BAB1	65	425427112503801	Co-op	112°50'39"	42°54'26"	1,339.77	0.030	446	2	1,333.26	0.003	0.03	1.28	2.61
06S 34E 09BCB1	66	425456112294001	Co-op	112°29'42"	42°54'54"	1,361.06	1.524	46	4	1,340.65	0.003	1.53	0.39	-7.66
06S 13E 08BDA2	67	425511114562301	Co-op	114°56'26"	42°55'10"	991.42	3.658	199	6	957.50	0.003	3.66	7.25	71.88
05S 33E 35CDC1	68	425608112340901	Co-op	112°34'11"	42°56'07"	1,349.64	0.003	1,770	6	1,341.09	0.003	0.01	0.35	1.79
05S 15E 35DBD2	69	425635114382302	Co-op	114°38'26"	42°56'34"	1,106.46	0.030	434	11	1,060.08	0.003	0.03	3.07	-10.15
05S 28E 26BBD1	70	425746113093901	Co-op	113°09'43"	42°57'45"	1,507.08	0.305	151	2	1,296.49	0.003	0.31	1.08	7.80

Table 5   55

Table 5. Wells in the Federal-State Cooperative and U.S. Geological Survey-Idaho National Laboratory water-level monitoring networks, eastern Snake River Plain, Idaho, during 2008.—Continued

Local name	Map No.	Site No.	Network name	Longitude	Latitude	Reference point elevation (m)	Reference point location error (m)	Sample size POR	Sample size 2008	2008 water-level elevation (m)	2008 measurement method error (m)	2008 measurement error (m)	Standard deviation (m)	Estimation error (m)
05S 17E 26ACA1	71	425746114240101	Co-op	114°24'07"	42°57'45"	1,211.72	0.003	2,616	6	1,141.81	0.003	0.01	3.37	-4.02
05S 31E 19DDC2	72	425754112521601	Co-op	112°52'15"	42°58'00"	1,348.25	3.048	34	2	1,332.45	0.003	3.05	0.77	-2.35
05S 31E 27ABA1	73	425757112485201	Co-op	112°48'57"	42°57'56"	1,342.10	0.030	2,253	3	1,334.65	0.003	0.03	1.21	-2.58
05S 25E 22DAD1	74	425812113271201	Co-op	113°27'15"	42°58'11"	1,397.98	0.003	178	4	1,240.29	0.003	0.01	2.51	2.18
05S 34E 20CBB2	75	425816112305102	Co-op	112°30'53"	42°58'15"	1,358.91	0.762	240	3	1,341.26	0.003	0.77	0.70	-1.71
05S 30E 12BBA1	76	430301112541301	Co-op	112°54'15"	43°00'32"	1,373.10	0.030	263	2	1,337.78	0.003	0.03	1.17	5.20
05S 14E 12AAA1	77	430400114435501	Co-op	114°43'58"	43°00'39"	1,100.90	3.048	187	6	1,061.87	0.003	3.05	2.19	8.98
04S 31E 36ABA1	78	430216112464001	Co-op	112°46'43"	43°02'18"	1,342.69	0.030	236	2	1,340.15	0.003	0.03	0.76	-1.71
04S 33E 20CBB1	79	430333112375801	Co-op	112°38'00"	43°03'32"	1,347.93	0.305	217	5	1,338.10	0.003	0.31	0.32	-3.87
04S 31E 20BBB1	80	430402112520301	Co-op	112°52'03"	43°04'00"	1,379.75	0.030	159	2	1,341.24	0.003	0.03	0.87	0.36
04S 31E 11ABA1	81	430547112473701	Co-op	112°47'40"	43°05'46"	1,361.96	1.524	152	3	1,341.97	0.003	1.53	0.73	-1.67
04S 32E 01CBA1	82	430607112400501	Co-op	112°40'07"	43°06'06"	1,357.38	1.524	225	5	1,343.78	0.003	1.53	0.99	2.46
04S 31E 05CBC1	83	430607112515601	Co-op	112°51'58"	43°06'06"	1,400.98	0.152	33	3	1,342.33	0.003	0.16	1.06	1.69
04S 33E 03CBB2	84	430610112353301	Co-op	112°35'36"	43°06'10"	1,356.75	0.030	1,400	3	1,343.87	0.003	0.03	0.69	1.08
04S 24E 06BBC1	85	430626113391001	Co-op	113°39'13"	43°06'25"	1,370.55	0.003	1,985	4	1,236.38	0.003	0.01	1.47	-8.28
03S 33E 25CCC1	86	430729112331201	Co-op	112°33'14"	43°07'28"	1,357.51	0.003	228	5	1,345.00	0.003	0.01	0.61	-0.23
03S 27E 24DDA1	87	430836113143401	Co-op	113°14'37"	43°08'35"	1,519.60	0.030	174	4	1,253.30	0.003	0.03	2.66	-44.93
03S 34E 22DAB1	88	430843112272701	Co-op	112°27'29"	43°08'42"	1,354.22	0.003	219	4	1,348.57	0.003	0.01	0.79	0.86
03S 31E 16CCB1	89	430930112505701	Co-op	112°51'01"	43°09'30"	1,415.31	1.524	126	2	1,343.38	0.003	1.53	0.79	2.27
03S 33E 17AAD1	90	430955112365001	Co-op	112°36'54"	43°09'57"	1,376.50	0.030	227	2	1,345.93	0.003	0.03	0.62	-0.25
03S 33E 14BBA1	91	431006112340901	Co-op	112°34'11"	43°10'05"	1,360.90	0.030	319	2	1,346.90	0.003	0.03	0.66	0.97
03S 34E 02BCC3	92	431126112271503	Co-op	112°27'17"	43°11'25"	1,356.18	0.305	232	5	1,347.77	0.003	0.31	0.87	-1.72
03S 32E 04ACA2	93	431138112425801	Co-op	112°43'00"	43°11'37"	1,383.29	1.524	5	2	1,346.69	0.003	1.53	1.24	2.37
02S 34E 33BBA1	94	431242112292801	Co-op	112°29'31"	43°12'40"	1,359.49	0.030	389	3	1,348.25	0.003	0.03	0.88	0.22
02S 35E 22DAC1	95	431349112202001	Co-op	112°20'22"	43°13'48"	1,380.28	0.762	160	3	1,352.95	0.003	0.77	1.40	-11.93
02S 35E 11DDD1	96	431517112190101	Co-op	112°19'03"	43°15'16"	1,378.27	0.003	229	5	1,369.68	0.003	0.01	4.02	10.03
02S 33E 16ABB1	97	431520112360901	Co-op	112°36'11"	43°15'19"	1,389.68	0.152	112	2	1,347.06	0.003	0.16	0.81	-0.68
02S 20E 01ACC2	98	431642114013002	Co-op	114°01'30"	43°16'46"	1,461.03	0.003	1,176	3	1,415.78	0.003	0.01	1.37	45.89
01S 35E 36CDD1	99	431702112182401	Co-op	112°18'24"	43°17'01"	1,384.86	1.524	14	3	1,365.92	0.003	1.53	0.41	-2.27
01S 23E 26CCC1	100	431810113413601	Co-op	113°41'42"	43°18'09"	1,534.17	0.305	193	4	1,236.73	0.003	0.31	2.72	-144.73
01S 34E 21DAC1	101	431902112284301	Co-op	112°28'45"	43°18'59"	1,386.96	1.524	124	3	1,350.81	0.003	1.53	1.04	-1.67
01S 32E 22BDB1	102	431929112421701	Co-op	112°42'20"	43°19'28"	1,445.77	1.524	125	2	1,346.82	0.003	1.53	0.90	-0.75
01S 22E 18DBD2	103	432007113525401	Co-op	113°52'54"	43°20'07"	1,468.65	3.048	16	3	1,443.30	0.003	3.05	2.66	95.31
02N 31E 35DCC1	104	432700112470801	Co-op, USGS-INL	112°47'11"	43°26'59"	1,531.86	0.003	2,033	14	1,350.69	0.003	0.01	1.10	-0.70
USGS 1														
02N 26E 22DDA1	105	432854113201001	Co-op	113°20'13"	43°28'53"	1,635.37	0.003	121	2	1,433.87	0.003	0.01	0.44	97.22
02N 38E 16ADD1	106	433029111590201	Co-op	111°59'04"	43°30'28"	1,445.22	0.762	147	2	1,408.74	0.003	0.77	3.13	-4.95
02N 37E 02ABA1	107	433220112040701	Co-op	112°04'09"	43°32'19"	1,441.24	0.003	678	9	1,391.24	0.003	0.01	2.46	-7.76
03N 38E 22BAB1	108	433457111583701	Co-op	111°58'40"	43°34'57"	1,461.07	1.524	196	3	1,421.90	0.003	1.53	4.26	-7.59

Table 5. Wells in the Federal-State Cooperative and U.S. Geological Survey-Idaho National Laboratory water-level monitoring networks, eastern Snake River Plain, Idaho, during 2008.—Continued

Local name	Map No.	Site No.	Network name	Longitude	Latitude	Reference point elevation (m)	Reference point location error (m)	Sample size POR	Sample size 2008	2008 water-level elevation (m)	2008 measurement method error (m)	2008 measurement error (m)	Standard deviation (m)	Estimation error (m)
03N 37E 12BDB1	109	433625112031801	Co-op	112°03'20"	43°36'24"	1,449.52	0.003	616	8	1,415.38	0.003	0.01	3.44	4.97
03N 37E 02CBD1	110	433656112043901	Co-op	112°04'41"	43°36'55"	1,458.02	3.048	209	2	1,407.96	0.003	3.05	3.12	-5.84
04N 39E 26DAA1	111	433849111492601	Co-op	111°49'28"	43°38'48"	1,501.32	0.610	292	6	1,485.97	0.003	0.61	7.06	24.63
04N 35E 14AAA1	112	434102112180701	Co-op, USGS-INL	112°18'09"	43°41'01"	1,506.59	0.003	922	6	1,376.63	0.003	0.01	1.62	-10.21
USBR SITE 15														
04N 38E 12BBB1	113	434153111563201	Co-op	111°56'34"	43°41'54"	1,473.14	0.003	233	3	1,464.76	0.003	0.01	2.64	3.07
05N 37E 21DBB1	114	434453112063601	Co-op	112°06'38"	43°44'52"	1,456.36	0.003	197	5	1,454.48	0.003	0.01	0.77	14.62
05N 39E 08DAD1	115	434638111530401	Co-op	111°53'06"	43°46'38"	1,473.39	0.003	327	3	1,471.07	0.003	0.01	0.37	0.09
05N 36E 02BDA2	116	434748112113602	Co-op	112°11'38"	43°47'47"	1,453.00	0.003	285	6	1,448.68	0.003	0.01	2.36	17.31
06N 35E 32DDD1	117	434756112212101	Co-op	112°21'23"	43°47'55"	1,460.78	0.003	199	6	1,381.10	0.003	0.01	2.01	-9.75
06N 39E 35CBB2	118	434816111501302	Co-op	111°50'15"	43°48'15"	1,476.50	0.003	309	5	1,474.19	0.003	0.01	0.90	2.32
06N 35E 27DDA1	119	434857112185801	Co-op	112°19'00"	43°48'56"	1,463.59	0.003	117	2	1,387.06	0.003	0.01	1.80	-24.97
06N 39E 30ADC1	120	434915111540501	Co-op	111°54'07"	43°49'14"	1,469.29	0.003	359	6	1,465.19	0.003	0.01	1.15	-1.78
06N 38E 25ACB4	121	434917111553102	Co-op	111°55'33"	43°49'16"	1,472.28	0.003	424	5	1,462.98	0.003	0.01	1.65	-1.10
06N 37E 29ACA2	122	434922112072202	Co-op	112°07'24"	43°49'21"	1,471.32	0.003	273	2	1,454.03	0.003	0.01	1.03	-2.98
06N 38E 30BAD2	123	434924112013801	Co-op	112°01'40"	43°49'23"	1,486.80	0.003	179	2	1,457.22	0.003	0.01	0.76	-3.89
USBR SITE 1														
06N 39E 28BBB1	124	434932111523701	Co-op	111°52'39"	43°49'31"	1,472.88	0.003	315	5	1,470.78	0.003	0.01	0.66	2.74
06N 39E 23AAC2	125	435015111495302	Co-op	111°49'55"	43°50'14"	1,477.49	0.003	567	3	1,464.39	0.003	0.01	1.29	-11.37
06N 35E 21AAB1	126	435028112202601	Co-op	112°20'28"	43°50'27"	1,459.41	0.003	372	2	1,427.75	0.003	0.01	1.69	22.06
06N 39E 16DAA1	127	435048111512701	Co-op	111°51'29"	43°50'47"	1,474.75	0.003	318	5	1,472.37	0.003	0.01	1.01	6.90
06N 40E 15AAA1	128	435115111430201	Co-op	111°43'04"	43°51'15"	1,494.60	3.048	105	6	1,488.70	0.003	3.05	1.42	-4.28
06N 39E 13ABA1	129	435118111481601	Co-op	111°48'18"	43°51'17"	1,483.48	0.003	276	5	1,480.29	0.003	0.01	1.39	10.37
06N 36E 11ABA3	130	435208112105103	Co-op	112°10'53"	43°52'07"	1,469.58	0.003	250	5	1,454.71	0.003	0.01	1.53	3.36
06N 39E 10BBB1	131	435209111512101	Co-op	111°51'21"	43°52'08"	1,474.55	0.003	390	5	1,464.22	0.003	0.01	1.55	-3.66
06N 38E 02DBD1	132	435228111563401	Co-op	111°56'36"	43°52'27"	1,489.95	0.003	199	5	1,463.07	0.003	0.01	1.68	0.10
07N 39E 34CCB1	133	435314111511902	Co-op	111°51'21"	43°53'13"	1,472.75	0.003	945	5	1,464.12	0.003	0.31	1.33	-4.98
07N 33E 34AAA1	134	435357112332001	Co-op	112°33'22"	43°53'56"	1,459.24	0.305	59	2	1,454.62	0.003	0.31	1.63	40.31
07N 35E 26CDD1	135	435359112182501	Co-op	112°18'27"	43°53'58"	1,461.08	0.305	154	6	1,453.82	0.003	0.31	1.85	6.31
07N 37E 28CCD1	136	435402112065001	Co-op	112°06'52"	43°54'01"	1,479.04	0.003	198	2	1,455.86	0.003	0.01	1.07	-1.94
07N 35E 20CBD1	137	435504112222301	Co-op	112°22'24"	43°55'04"	1,469.67	0.003	2,909	6	1,451.60	0.003	0.01	2.38	1.06
07N 38E 23DBA2	138	435506111563102	Co-op	111°56'34"	43°55'05"	1,480.10	0.003	2,586	3	1,465.33	0.003	0.01	0.91	3.30
07N 40E 19ADD4	139	435516111464002	Co-op	111°46'42"	43°55'15"	1,481.46	0.003	763	5	1,480.24	0.003	0.01	0.76	-9.71
07N 36E 22ABD4	140	435528112121201	Co-op	112°12'12"	43°55'27"	1,461.61	0.003	1,171	6	1,455.53	0.003	0.01	0.85	-0.19
07N 34E 24BBA1	141	435540112243901	Co-op	112°24'41"	43°55'39"	1,461.09	1.524	127	6	1,450.96	0.003	1.53	3.07	0.95
07N 39E 16DBB3	142	435605111515801	Co-op	111°52'00"	43°56'04"	1,486.23	0.003	1,063	5	1,480.36	0.003	0.01	2.15	12.14
07N 35E 13AAD1	143	435626112164301	Co-op	112°16'45"	43°56'25"	1,460.93	0.003	247	5	1,454.90	0.003	0.01	1.69	-1.41
07N 39E 07BDA1	144	435705111542701	Co-op	111°54'29"	43°57'04"	1,486.83	0.003	590	5	1,463.96	0.003	0.01	1.85	-6.63
07N 36E 11ABB1	145	435723112111101	Co-op	112°11'10"	43°57'23"	1,496.14	1.524	22	6	1,456.71	0.003	1.53	0.53	0.45

Table 5   57

Table 5. Wells in the Federal-State Cooperative and U.S. Geological Survey-Idaho National Laboratory water-level monitoring networks, eastern Snake River Plain, Idaho, during 2008.—Continued

Local name	Map No.	Site No.	Network name	Longitude	Latitude	Reference point elevation (m)	Reference point location error (m)	Sample size POR	Sample size 2008	2008 water-level elevation (m)	2008 measurement method error (m)	2008 measurement error (m)	Standard deviation (m)	Estimation error (m)
07N 36E 09BBB1	146	435728112141301	Co-op	112°14'15"	43°57'27"	1,462.60	1.524	133	6	1,456.29	0.003	1.53	1.29	0.31
07N 34E 04CDC1	147	435728112281101	Co-op	112°28'11"	43°57'27"	1,461.62	0.003	1,175	6	1,452.08	0.003	0.01	2.36	0.29
07N 40E 05DBC1	148	435736111460201	Co-op	111°46'04"	43°57'35"	1,500.64	0.003	371	5	1,497.55	0.003	0.01	0.99	20.13
08N 41E 33ABB1	149	435904111373101	Co-op	111°37'35"	43°59'04"	1,528.11	1.524	178	6	1,508.19	0.003	1.53	2.57	0.13
08N 34E 27DD1	150	435912112264801	Co-op	112°26'50"	43°59'11"	1,465.66	3.048	156	6	1,449.94	0.003	3.05	2.84	-5.01
08N 41E 25CBB2	151	435924111343702	Co-op	111°34'39"	43°59'23"	1,547.93	1.524	24	6	1,527.00	0.003	1.53	3.04	8.59
08N 36E 21DCD1	152	440002112131801	Co-op	112°13'20"	44°00'01"	1,467.18	0.762	157	6	1,456.59	0.003	0.77	1.39	-0.18
08N 34E 17CCC7	153	440058112293605	Co-op	112°29'38"	44°00'57"	1,466.85	0.003	230	3	1,456.10	0.003	0.01	1.07	5.83
08N 34E 11DCC1	154	440151112252301	Co-op	112°25'25"	44°01'50"	1,485.48	1.524	158	6	1,455.76	0.003	1.53	1.56	2.94
08N 40E 06CCC1	155	440236111474701	Co-op	111°47'49"	44°02'35"	1,552.50	1.524	133	6	1,464.86	0.003	1.53	1.59	0.36
08N 36E 03DCD1	156	440239112121101	Co-op	112°12'13"	44°02'38"	1,477.85	0.762	106	6	1,458.18	0.003	0.77	1.45	2.59
08N 40E 01CAD1	157	440253111412101	Co-op	111°41'23"	44°02'52"	1,574.15	0.003	207	3	1,467.93	0.003	0.01	2.71	-23.57
09N 36E 33CBB1	158	440353112135701	Co-op	112°13'59"	44°03'52"	1,483.96	1.524	1,416	6	1,456.08	0.003	1.53	1.34	-1.24
09N 34E 29DAB1	159	440447112284401	Co-op	112°28'47"	44°04'46"	1,475.72	0.762	64	3	1,452.72	0.003	0.77	2.92	-3.01
09N 36E 15CCC1	160	440608112125001	Co-op	112°12'52"	44°06'07"	1,510.49	1.524	148	4	1,458.53	0.003	1.53	1.59	3.09
09N 34E 11ADD1	161	440725112245301	Co-op	112°24'55"	44°07'24"	1,508.35	1.524	1,638	6	1,456.59	0.003	1.53	0.98	-3.94
09N 40E 05DDD1	162	440752111452901	Co-op	111°45'31"	44°07'51"	1,688.31	0.003	185	4	1,458.04	0.003	0.01	2.49	-61.99
09N 38E 05BBA1	163	440839112003101	Co-op	112°00'33"	44°08'38"	1,676.04	1.524	133	6	1,458.43	0.003	1.53	1.38	-66.97
09N 36E 04BAA1	164	440841112213001	Co-op	112°13'30"	44°08'40"	1,541.89	1.524	154	6	1,457.48	0.003	1.53	1.73	0.36
10N 36E 21CCC1	165	441030112135801	Co-op	112°14'06"	44°10'27"	1,567.80	1.524	90	6	1,457.33	0.003	1.53	1.44	-19.16
11N 39E 07DBC1 18 MILE RANCH	166	441740111540201	Co-op	111°54'04"	44°17'39"	1,904.40	1.524	130	6	1,740.53	0.003	1.53	2.88	267.52
01S 30E 15BCA1 USGS 14 MV-61	167	432019112563201	USGS-INL	112°56'34"	43°20'18"	1,565.56	0.003	931	5	1,345.81	0.003	0.01	0.68	3.50
01N 30E 29CCB1 USGS 124	168	432307112583101	USGS-INL	112°58'31"	43°23'06"	1,556.25	0.003	42	2	1,346.56	0.003	0.01	0.74	0.08
01N 29E 30BBD1 USGS 11	169	432336113064201	USGS-INL	113°06'45"	43°23'35"	1,545.51	0.003	540	5	1,344.92	0.003	0.01	0.86	8.59
01N 29E 08BCD1 USGS 125	170	432602113052801	USGS-INL	113°05'33"	43°25'59"	1,540.51	0.003	67	4	1,347.10	0.003	0.01	0.84	1.36
01N 30E 10BBA1 CERRO GRANDE	171	432618112555501	USGS-INL	112°55'56"	43°26'17"	1,518.73	0.003	437	4	1,347.85	0.003	0.01	1.21	-0.23
02N 29E 35CCC1 USGS 108	172	432659112582601	USGS-INL	112°58'29"	43°26'58"	1,534.61	0.003	67	1	1,348.14	0.003	0.01	1.21	0.09
02N 29E 31CDC1 USGS 109	173	432701113025601	USGS-INL	113°02'58"	43°27'00"	1,538.36	0.003	131	4	1,346.92	0.003	0.01	1.12	-0.82
02N 29E 33DCC1 USGS 105	174	432703113001801	USGS-INL	113°00'20"	43°27'03"	1,554.05	0.003	69	3	1,348.05	0.003	0.01	1.27	0.29

Table 5. Wells in the Federal-State Cooperative and U.S. Geological Survey-Idaho National Laboratory water-level monitoring networks, eastern Snake River Plain, Idaho, during 2008.—Continued

Map No.	Local name	Site No.	Network name	Longitude	Latitude	Reference point elevation (m)	Reference point location error (m)	Sample size POR	Sample size 2008	2008 water-level elevation (m)	2008 measurement method error (m)	2008 measurement error (m)	Standard deviation (m)	Estimation error (m)
175	02N 30E 35DAD2 USGS 110A	432717112501502	USGS-INL	112°50'17"	43°27'16"	1,524.88	0.003	37	2	1,350.85	0.003	0.01	0.92	0.59
176	02N 27E 33ACC2 USGS 13	432731113143902	USGS-INL	113°14'41"	43°27'30"	1,639.25	0.003	376	1	1,337.06	0.003	0.01	0.39	-39.02
177	02N 28E 35AAC1 USGS 9	432740113044501	USGS-INL	113°04'42"	43°27'32"	1,534.29	0.003	1,885	12	1,347.24	0.003	0.01	0.99	0.05
178	02N 29E 19DDA1 A11A31	432853113021701	USGS-INL	113°02'19"	43°28'52"	1,545.04	0.003	14	2	1,348.00	0.003	0.01	0.91	-0.19
179	02N 29E 24DAD1 USGS 104	432856112560801	USGS-INL	112°56'11"	43°28'55"	1,521.28	0.003	123	5	1,349.84	0.003	0.01	1.08	-0.22
180	02N 29E 19BCB1 USGS 120	432919113031501	USGS-INL	113°03'17"	43°29'18"	1,537.38	0.003	329	12	1,348.21	0.003	0.01	0.94	-0.30
181	02N 32E 22ABA1 COREHOLE 1	432927112410101	USGS-INL	112°41'03"	43°29'26"	1,637.88	0.003	162	4	1,350.69	0.003	0.01	1.18	-6.27
182	02N 29E 20BBA1 RWMC M6S	432931113015001	USGS-INL	113°01'53"	43°29'30"	1,545.20	0.003	28	1	1,348.68	0.003	0.01	0.74	0.31
183	02N 28E 21BBB1 SGS 86	432935113080001	USGS-INL	113°08'04"	43°29'34"	1,548.50	0.003	340	5	1,348.48	0.003	0.01	1.38	4.35
184	02N 29E 18CCD2 RWMC M4D	432939113030101	USGS-INL	113°03'04"	43°29'38"	1,532.05	0.003	29	1	1,348.87	0.003	0.01	0.71	0.11
185	02N 29E 18CCD1 USGS 88	432940113030201	USGS-INL	113°03'04"	43°29'39"	1,531.40	0.003	308	5	1,348.78	0.003	0.01	4.67	-0.06
186	02N 30E 16CCA1 USGS 107	432942112532801	USGS-INL	112°53'30"	43°29'41"	1,499.90	0.003	115	2	1,351.58	0.003	0.01	1.14	-1.47
187	02N 29E 18DCB1 USGS 119	432945113023401	USGS-INL	113°02'36"	43°29'44"	1,534.77	0.003	119	4	1,348.61	0.003	0.01	0.96	-0.17
188	02N 29E 18DCA1 USGS 118	432947113023001	USGS-INL	113°02'33"	43°29'46"	1,528.85	0.003	72	4	1,348.79	0.003	0.01	0.84	0.15
189	02N 29E 18CBD1 USGS 117	432955113025901	USGS-INL	113°03'01"	43°29'54"	1,528.87	0.003	127	4	1,348.77	0.003	0.01	0.93	0.10
190	02N 29E 18CBA1 RWMC M1SA	432956113030901	USGS-INL	113°03'11"	43°29'55"	1,528.52	0.003	40	2	1,348.50	0.003	0.01	0.73	-0.33
191	02N 29E 15CBA1 USGS 106	432959112593101	USGS-INL	112°59'34"	43°29'58"	1,529.72	0.003	123	2	1,348.61	0.003	0.01	1.38	-0.60
192	02N 28E 13ADD1 USGS 89	433005113032801	USGS-INL	113°03'34"	43°30'05"	1,534.28	0.003	297	4	1,348.98	0.003	0.01	2.27	0.37
193	02N 29E 18ADB1 RWMC M3S	433008113021801	USGS-INL	113°02'21"	43°30'07"	1,530.04	0.003	48	2	1,348.75	0.003	0.01	0.90	-0.23

Table 5   59

Table 5. Wells in the Federal-State Cooperative and U.S. Geological Survey-Idaho National Laboratory water-level monitoring networks, eastern Snake River Plain, Idaho, during 2008.—Continued

Local name	Map No.	Site No.	Network name	Longitude	Latitude	Reference point elevation (m)	Reference point location error (m)	Sample size POR	Sample size 2008	2008 water-level elevation (m)	2008 measurement method error (m)	2008 measurement error (m)	Standard deviation (m)	Estimation error (m)
02N 29E 18BDA1 USGS 87	194	433013113024201	USGS-INL	113°02'46"	43°30'12"	1,530.50	0.003	303	4	1,349.11	0.003	0.01	1.41	0.33
02N 29E 13AAA1 USGS 83	195	433023112561501	USGS-INL	112°56'18"	43°30'22"	1,507.10	0.003	359	4	1,352.95	0.003	0.01	1.04	-0.07
02N 29E 17BBA1 RWMC M7S	196	433023113014801	USGS-INL	113°01'51"	43°30'22"	1,526.63	0.003	46	2	1,348.96	0.003	0.01	2.43	0.25
02N 29E 11CCA1 USGS 131	197	433036112581601	USGS-INL	112°58'19"	43°30'35"	1,518.14	0.003	107	12	1,351.36	0.003	0.01	0.20	-0.98
02N 29E 09CDA2 USGS 129	198	433036113002701	USGS-INL	113°00'30"	43°30'36"	1,533.04	0.003	89	11	1,348.57	0.003	0.01	0.15	-0.91
02N 29E 07DAA1 RWMC M14S	199	433052113025001	USGS-INL	113°02'52"	43°30'51"	1,534.96	0.003	27	2	1,348.89	0.003	0.01	1.18	-0.52
02N 29E 11ADD1 USGS 127	200	433058112572201	USGS-INL	112°57'25"	43°30'57"	1,511.78	0.003	99	4	1,354.78	0.003	0.01	0.97	1.42
02N 29E 08ADC1 RWMC M11S	201	433058113010401	USGS-INL	113°01'06"	43°30'57"	1,523.29	0.003	24	2	1,349.30	0.003	0.01	0.66	-0.05
02N 29E 03CCC1 WMC M12S	202	433118112593401	USGS-INL	112°59'36"	43°31'17"	1,517.51	0.003	25	2	1,352.50	0.003	0.01	0.81	1.35
02N 27E 02DDC1 USGS 8	203	433121113115801	USGS-INL	113°12'00"	43°31'20"	1,584.48	0.003	519	5	1,348.70	0.003	0.01	1.24	-17.60
02N 30E 04DCC1 SITE 9	204	433123112530101	USGS-INL	112°53'03"	43°31'22"	1,502.39	0.003	355	4	1,355.85	0.003	0.01	1.21	0.94
02N 29E 01DCA1 USGS 130	205	433130112562801	USGS-INL	112°56'31"	43°31'30"	1,502.97	0.003	113	12	1,355.36	0.003	0.01	0.29	0.15
03N 29E 36DDC1 ICPP 1798	206	433216112562601	USGS-INL	112°56'46"	43°32'26"	1,504.13	0.003	56	6	1,356.36	0.003	0.01	0.23	0.35
03N 29E 36DCC2 CFALF 2-10	207	433216112563301	USGS-INL	112°56'35"	43°32'15"	1,504.34	0.003	76	5	1,355.51	0.003	0.01	1.28	-0.72
03N 29E 36CCC1 CFALF 3-9	208	433216112571001	USGS-INL	112°57'13"	43°32'15"	1,507.10	0.003	27	2	1,356.33	0.003	0.01	1.40	0.38
02N 35E 02BBC1 HIGHWAY 1C	209	433218112191601	USGS-INL	112°19'18"	43°32'17"	1,552.49	0.003	430	4	1,372.09	0.003	0.01	1.43	3.44
03N 31E 35DCA1 AREA 2	210	433223112470201	USGS-INL	112°47'05"	43°32'22"	1,564.27	0.003	75	2	1,357.49	0.003	0.01	1.43	0.63
03N 29E 36DAC1 CFALF 2-11	211	433230112561701	USGS-INL	112°56'20"	43°32'30"	1,503.23	0.003	21	2	1,356.27	0.003	0.01	1.84	0.33
03N 29E 36BCB1 USGS 85	212	433246112571201	USGS-INL	112°57'14"	43°32'45"	1,506.47	0.003	362	4	1,356.33	0.003	0.01	1.46	-0.05
03N 29E 36BDB3 USGS 128	213	433250112565601	USGS-INL	112°56'58"	43°32'49"	1,505.22	0.003	45	6	1,356.28	0.003	0.01	0.76	-0.12

Table 5. Wells in the Federal-State Cooperative and U.S. Geological Survey-Idaho National Laboratory water-level monitoring networks, eastern Snake River Plain, Idaho, during 2008.—Continued

Local name	Map No.	Site No.	Network name	Longitude	Latitude	Reference point elevation (m)	Reference point location error (m)	Sample size POR	Sample size 2008	2008 water-level elevation (m)	2008 measurement method error (m)	2008 measurement error (m)	Standard deviation (m)	Estimation error (m)
03N 30E 31AAD1 USGS 20	214	433253112545901	USGS-INL	112°55'02"	43°32'52"	1,499.18	0.003	715	4	1,355.91	0.003	0.01	1.12	-0.67
03N 32E 36ADD1 USGS 101	215	433255112381801	USGS-INL	112°38'22"	43°32'55"	1,601.64	0.003	180	11	1,363.97	0.003	0.01	1.54	3.98
03N 29E 34ADD1 ICPP MON A-166	216	433300112583301	USGS-INL	112°58'36"	43°32'59"	1,511.69	0.003	58	4	1,356.37	0.003	0.01	0.38	0.68
03N 34E 32BBC1 HIGHWAY 2	217	433307112300001	USGS-INL	112°30'02"	43°33'06"	1,591.10	0.015	385	4	1,367.32	0.003	0.02	1.34	4.03
03N 29E 25DDB1 USGS 113	218	433314112561801	USGS-INL	112°56'21"	43°33'14"	1,502.30	0.003	107	4	1,356.74	0.003	0.01	1.68	0.49
03N 29E 25DCA1 USGS 112	219	433314112563001	USGS-INL	112°56'33"	43°33'14"	1,503.06	0.003	114	5	1,356.29	0.003	0.01	1.67	-0.24
03N 30E 30CCB1 USGS 77	220	433315112560301	USGS-INL	112°56'06"	43°33'14"	1,501.11	0.003	106	2	1,356.24	0.003	0.01	1.59	-0.24
03N 30E 30CBD1 USGS 114	221	433318112555001	USGS-INL	112°55'53"	43°33'18"	1,500.69	0.003	106	4	1,356.34	0.003	0.01	1.70	-0.01
03N 32E 29DDC1 USGS 2	222	433320112432301	USGS-INL	112°43'24"	43°33'19"	1,563.25	0.003	685	5	1,359.42	0.003	0.01	1.08	0.23
03N 30E 30CAD1 USGS 115	223	433320112554101	USGS-INL	112°55'44"	43°33'19"	1,500.33	0.003	109	3	1,356.47	0.003	0.01	1.63	0.12
03N 29E 25CAD1 USGS 38	224	433322112564301	USGS-INL	112°56'46"	43°33'22"	1,503.43	0.003	140	3	1,356.26	0.003	0.01	1.63	-0.02
03N 29E 25CAA1 USGS 37	225	433326112564801	USGS-INL	112°56'51"	43°33'25"	1,503.28	0.003	343	2	1,356.27	0.003	0.01	1.24	-0.07
03N 29E 25BDD1 USGS 36	226	433330112565201	USGS-INL	112°56'54"	43°33'29"	1,503.37	0.003	174	4	1,356.41	0.003	0.01	1.75	0.07
03N 30E 30ACC1 USGS 116	227	433331112553201	USGS-INL	112°55'35"	43°33'31"	1,499.47	0.003	105	4	1,356.49	0.003	0.01	1.68	0.06
03N 30E 30BCC1 USGS 111	228	433331112560501	USGS-INL	112°56'08"	43°33'30"	1,500.82	0.003	71	2	1,356.16	0.003	0.01	1.81	-0.29
03N 29E 26CAB1 ICPP MON A-167	229	433331112580701	USGS-INL	112°58'09"	43°33'30"	1,508.84	0.003	40	2	1,356.56	0.003	0.01	0.47	0.13
03N 29E 25BDC1 USGS 34	230	433334112565501	USGS-INL	112°56'57"	43°33'34"	1,503.28	0.003	166	2	1,356.40	0.003	0.01	1.59	0.07
03N 29E 25BDB1 USGS 35	231	433339112565801	USGS-INL	112°57'01"	43°33'38"	1,503.41	0.003	127	2	1,356.24	0.003	0.01	1.67	-0.17
03N 29E 25BBD1 USGS 39	232	433343112570001	USGS-INL	112°57'04"	43°33'42"	1,503.89	0.003	177	4	1,356.42	0.003	0.01	1.77	0.16

Table 5   61

Table 5. Wells in the Federal-State Cooperative and U.S. Geological Survey-Idaho National Laboratory water-level monitoring networks, eastern Snake River Plain, Idaho, during 2008.—Continued

Local name	Map No.	Site No.	Network name	Longitude	Latitude	Reference point elevation (m)	Reference point location error (m)	Sample size POR	Sample size 2008	2008 water-level elevation (m)	2008 measurement method error (m)	2008 measurement error (m)	Standard deviation (m)	Estimation error (m)
03N 30E 30BAD1 USGS 67	233	433344112554101	USGS-INL	112°55'43"	43°33'43"	1,498.65	0.003	79	2	1,356.41	0.003	0.01	1.84	-0.07
03N 29E 25ABD1 USGS 57	234	433344112562601	USGS-INL	112°56'28"	43°33'43"	1,501.36	0.003	138	4	1,356.42	0.003	0.01	1.73	0.15
03N 30E 30BBB1 USGS 51	235	433350112560601	USGS-INL	112°56'09"	43°33'49"	1,499.74	0.003	87	2	1,356.59	0.003	0.01	1.81	0.25
03N 29E 25AAA2 USGS 123	236	433352112561401	USGS-INL	112°56'16"	43°33'51"	1,500.45	0.003	46	2	1,356.30	0.003	0.01	1.43	-0.17
03N 30E 30BBA2 USGS 122	237	433353112555201	USGS-INL	112°55'54"	43°33'53"	1,498.79	0.003	44	2	1,356.46	0.003	0.01	1.46	-0.07
03N 30E 30BAB1 USGS 59	238	433354112554701	USGS-INL	112°55'51"	43°33'53"	1,498.51	0.003	81	2	1,356.53	0.003	0.01	1.80	0.07
03N 29E 23DCD1 USGS 84	239	433356112574201	USGS-INL	112°57'44"	43°33'56"	1,506.05	0.003	333	4	1,356.46	0.003	0.01	1.53	0.09
03N 30E 19DDC2 USGS 82	240	433401112551001	USGS-INL	112°55'13"	43°34'00"	1,496.66	0.003	350	4	1,356.65	0.003	0.01	1.52	-0.22
03N 30E 19CCC1 USGS 48	241	433401112560301	USGS-INL	112°56'05"	43°34'00"	1,499.72	0.003	101	2	1,356.52	0.003	0.01	1.95	-0.01
03N 29E 24DDC1 USGS 45	242	433402112561801	USGS-INL	112°56'20"	43°34'02"	1,499.97	0.003	102	2	1,356.25	0.003	0.01	1.93	-0.22
03N 29E 24DDA2 USGS 42	243	433404112561301	USGS-INL	112°56'14"	43°34'02"	1,499.72	0.003	110	2	1,356.48	0.003	0.01	1.84	0.05
03N 30E 19CCB1 USGS 47	244	433407112560301	USGS-INL	112°56'06"	43°34'07"	1,499.35	0.003	114	2	1,356.58	0.003	0.01	1.77	0.05
03N 29E 24DDA3 SGS 46	245	433407112561501	USGS-INL	112°56'17"	43°34'06"	1,499.59	0.003	304	2	1,356.56	0.003	0.01	1.36	0.09
03N 29E 24DDA1 USGS 41	246	433409112561301	USGS-INL	112°56'14"	43°34'07"	1,499.54	0.003	117	2	1,356.52	0.003	0.01	1.85	-0.05
03N 29E 24DDB1 USGS 44	247	433409112562101	USGS-INL	112°56'24"	43°34'08"	1,499.99	0.003	109	2	1,356.50	0.003	0.01	1.86	0.08
03N 29E 24DAD1 USGS 40	248	433411112561101	USGS-INL	112°56'14"	43°34'11"	1,499.35	0.003	352	4	1,356.54	0.003	0.01	1.60	0.02
03N 30E 19CAC1 USGS 52	249	433414112554201	USGS-INL	112°55'47"	43°34'14"	1,497.46	0.003	83	2	1,356.53	0.003	0.01	1.81	-0.07
03N 29E 24DAD2 USGS 43	250	433415112561501	USGS-INL	112°56'17"	43°34'14"	1,499.30	0.003	117	2	1,356.50	0.003	0.01	1.89	-0.01
03N 29E 23CBA1 MIDDLE 1823	251	433418112581701	USGS-INL	112°58'20"	43°34'18"	1,506.58	0.003	60	11	1,356.42	0.003	0.01	0.16	0.00

Table 5. Wells in the Federal-State Cooperative and U.S. Geological Survey-Idaho National Laboratory water-level monitoring networks, eastern Snake River Plain, Idaho, during 2008.—Continued

Map No.	Local name	Site No.	Network name	Longitude	Latitude	Reference point elevation (m)	Reference point location error (m)	Sample size POR	Sample size 2008	2008 water-level elevation (m)	2008 measurement method error (m)	2008 measurement error (m)	Standard deviation (m)	Estimation error (m)
252	03N 29E 19CBB1 USGS 22	43342211303701	USGS-INL	113°03'24"	43°34'21"	1,539.78	0.003	593	4	1,351.45	0.003	0.01	1.27	-4.96
253	03N 29E 23ADC1 USGS 76	43342511257320 1	USGS-INL	112°57'35"	43°34'24"	1,503.65	0.003	102	3	1,356.17	0.003	0.01	1.74	-0.70
254	03N 29E 23ABB1 USGS 65	43344711257450 1	USGS-INL	112°57'50"	43°34'46"	1,502.13	0.003	228	4	1,357.49	0.003	0.01	1.46	1.18
255	03N 30E 16DDD1 NPR TEST	43344911252310 1	USGS-INL	112°52'34"	43°34'49"	1,504.68	0.003	125	5	1,360.21	0.003	0.01	1.94	1.47
256	03N 30E 18CCC1 USGS 121	43345011256030 1	USGS-INL	112°56'06"	43°34'49"	1,497.53	0.003	49	3	1,356.31	0.003	0.01	1.42	-1.51
257	03N 29E 14DDA2 USGS 58	43350011257250 2	USGS-INL	112°57'28"	43°24'59"	1,500.11	0.003	415	4	1,356.67	0.003	0.01	1.45	-0.10
258	03N 32E 14CDD1 USGS 100	43350311240070 1	USGS-INL	112°40'09"	43°35'02"	1,573.23	0.003	130	4	1,364.18	0.003	0.01	1.61	0.11
259	03N 29E 14CBD1 USGS 79	43350511258190 1	USGS-INL	112°58'22"	43°35'05"	1,503.89	0.003	288	3	1,356.25	0.003	0.01	1.40	-0.45
260	03N 29E 14DBD1 TRA DISP	43350611257230 1	USGS-INL	112°57'39"	43°35'05"	1,501.49	0.003	63	2	1,356.52	0.003	0.01	1.68	-0.21
261	03N 32E 13DCA1 ARBOR TEST	43350911238480 1	USGS-INL	112°38'50"	43°35'08"	1,575.06	0.003	387	4	1,364.99	0.003	0.01	1.32	-0.24
262	03N 29E 14ADD1 MTR TEST	43352011257260 1	USGS-INL	112°57'32"	43°35'19"	1,499.61	0.003	2,678	12	1,356.67	0.003	0.01	1.25	-0.55
263	03N 29E 14BCB1 SITE 19	43352211258210 1	USGS-INL	112°58'24"	43°35'21"	1,502.50	0.003	87	2	1,356.67	0.003	0.01	1.73	0.13
264	03N 30E 12CDD1 USGS 5	43354311249380 1	USGS-INL	112°49'40"	43°35'42"	1,506.03	0.003	441	4	1,360.29	0.003	0.01	4.44	-0.09
265	03N 32E 13BBD1 SITE 16	43354511239150 1	USGS-INL	112°39'18"	43°35'44"	1,562.04	0.003	106	1	1,365.44	0.003	0.01	1.15	0.04
266	03N 32E 14AAC1 ANL OBS A 001	43354511239410 1	USGS-INL	112°39'44"	43°35'44"	1,561.98	0.003	16	1	1,365.34	0.003	0.01	1.34	0.15
267	03N 32E 14AAC2 ANL MW 13	43354511239410 2	USGS-INL	112°39'44"	43°35'45"	1,561.77	0.003	16	1	1,365.20	0.003	0.01	1.22	-0.14
268	03N 29E 12DDB1 FIRE STA 2	43354811256230 1	USGS-INL	112°56'25"	43°35'47"	1,495.29	0.003	41	5	1,361.21	0.003	0.01	3.00	3.28
269	03N 29E 01DBB1 USGS 98	43365711256360 1	USGS-INL	112°56'38"	43°36'56"	1,489.30	0.003	162	3	1,359.86	0.003	0.01	2.45	-1.18
270	03N 30E 06ACD1 USGS 99	43370511255210 1	USGS-INL	112°55'24"	43°37'03"	1,485.91	0.003	157	4	1,361.49	0.003	0.01	2.31	0.42

Table 5   63

Table 5. Wells in the Federal-State Cooperative and U.S. Geological Survey-Idaho National Laboratory water-level monitoring networks, eastern Snake River Plain, Idaho, during 2008.—Continued

Local name	Map No.	Site No.	Network name	Longitude	Latitude	Reference point elevation (m)	Reference point location error (m)	Sample size POR	Sample size 2008	2008 water-level elevation (m)	2008 measurement method error (m)	2008 measurement error (m)	Standard deviation (m)	Estimation error (m)
03N 29E 01ABC1 WS INEL 1	271	4337161112563601	USGS-INL	112°56'42"	43°37'14"	1,486.12	0.003	82	3	1,361.10	0.003	0.01	2.38	1.01
04N 35E 31DAA1	272	4337591122225401	USGS-INL	112°22'56"	43°37'58"	1,561.68	1.524	17	1	1,373.38	0.003	1.53	1.44	-0.39
04N 30E 31ABD1 USGS 97	273	4338071125515501	USGS-INL	112°55'19"	43°38'06"	1,481.93	0.003	446	12	1,361.64	0.003	0.01	2.48	-0.06
04N 30E 26CCA1 SITE 6	274	4338261125510701	USGS-INL	112°51'09"	43°38'25"	1,475.13	0.003	53	2	1,361.98	0.003	0.01	2.26	0.53
04N 30E 30DAB1 NRF 9	275	4338401125550201	USGS-INL	112°55'03"	43°38'35"	1,480.40	0.003	53	2	1,361.65	0.003	0.01	2.07	-0.25
04N 30E 29CBB1 NRF 10	276	4338411125452201	USGS-INL	112°54'51"	43°38'37"	1,480.57	0.003	52	2	1,361.92	0.003	0.01	2.07	0.07
04N 30E 30ADC1 NRF 8	277	4338431125550901	USGS-INL	112°55'14"	43°38'41"	1,480.33	0.003	52	2	1,361.92	0.003	0.01	2.07	0.18
04N 30E 29BCD1 NRF 11	278	4338471125544201	USGS-INL	112°54'42"	43°38'43"	1,479.96	3.048	51	2	1,362.03	0.003	3.05	2.11	0.20
04N 30E 30ACA1 USGS 102	279	4338531125551601	USGS-INL	112°55'19"	43°38'50"	1,479.43	0.003	108	3	1,361.86	0.003	0.01	1.96	0.09
04N 30E 29BAC1 NRF 12	280	4338551125543201	USGS-INL	112°54'30"	43°38'53"	1,479.60	0.003	52	2	1,361.70	0.003	0.01	2.02	-0.17
04N 30E 19DDD1 NRF 6	281	4339101125550101	USGS-INL	112°55'04"	43°39'10"	1,478.21	0.003	74	3	1,361.36	0.003	0.01	1.92	-0.42
04N 30E 20CCA1 NRF 7	282	4339201125543601	USGS-INL	112°54'33"	43°39'23"	1,477.11	0.003	70	2	1,361.72	0.003	0.01	1.79	0.06
04N 30E 19DAD1 NRF 13	283	4339281125545401	USGS-INL	112°54'52"	43°39'32"	1,476.91	3.048	52	3	1,361.69	0.003	3.05	2.10	-0.11
04N 30E 22BDD1 USGS 17	284	4339371125515401	USGS-INL	112°51'57"	43°39'36"	1,474.30	0.003	601	6	1,362.18	0.003	0.01	1.71	-0.58
04N 35E 20CAA1	285	4339451122221701	USGS-INL	112°22'19"	43°39'45"	1,534.25	1.524	22	3	1,376.35	0.003	1.53	1.72	3.69
04N 29E 14CAA1 SITE 17	286	4340271125775701	USGS-INL	112°57'59"	43°40'26"	1,488.64	0.003	173	12	1,363.27	0.003	0.01	2.50	0.35
04N 31E 16ADC1 USGS 6	287	4340311124537011	USGS-INL	112°45'39"	43°40'30"	1,494.15	0.003	472	4	1,364.95	0.003	0.01	1.24	-3.24
04N 29E 09DCD1 USGS 23	288	4340551125959901	USGS-INL	113°00'03"	43°40'54"	1,489.77	0.003	556	5	1,363.48	0.003	0.01	1.97	-3.30
04N 30E 07ADB1 USGS 12	289	4341261125507011	USGS-INL	112°55'10"	43°41'25"	1,469.90	0.003	2,207	12	1,365.26	0.003	0.01	2.10	1.44
04N 30E 06ABA1 USGS 15	290	4342341125511701	USGS-INL	112°55'20"	43°42'34"	1,467.77	0.003	220	4	1,365.88	0.003	0.01	2.47	-2.95

Table 5. Wells in the Federal-State Cooperative and U.S. Geological Survey-Idaho National Laboratory water-level monitoring networks, eastern Snake River Plain, Idaho, during 2008.—Continued

Local name	Map No.	Site No.	Network name	Longitude	Latitude	Reference point elevation (m)	Reference point location error (m)	Sample size POR	Sample size 2008	2008 water-level elevation (m)	2008 measurement method error (m)	2008 measurement error (m)	Standard deviation (m)	Estimation error (m)
05N 32E 36ADD1 USGS 21	291	434307112382601	USGS-INL	112°38'28"	43°43'07"	1,475.92	0.003	3,549	63	1,371.37	0.003	0.01	1.58	-2.53
05N 31E 28CCC1 SITE 14	292	434334112463101	USGS-INL	112°46'34"	43°43'34"	1,462.13	0.003	1,097	4	1,375.41	0.003	0.01	1.51	4.23
05N 34E 29DAA1 USGS 29	293	434407112285101	USGS-INL	112°28'53"	43°44'06"	1,487.75	0.003	144	2	1,376.08	0.003	0.01	1.70	3.77
05N 29E 23CDD1 USGS 19	294	434426112575701	USGS-INL	112°57'59"	43°44'26"	1,464.13	0.003	2,653	11	1,378.15	0.003	0.01	1.32	7.58
05N 33E 23DDA1 USGS 32	295	434444112322101	USGS-INL	112°32'24"	43°44'43"	1,467.79	0.003	122	2	1,375.63	0.003	0.01	1.79	0.79
05N 31E 14BCC1 USGS 18	296	434540112440901	USGS-INL	112°44'12"	43°45'40"	1,465.39	0.003	528	5	1,378.29	0.003	0.01	1.77	-1.40
05N 30E 15ADC1 DH 2A	297	434547112512801	USGS-INL	112°51'30"	43°45'47"	1,462.46	0.003	123	4	1,374.44	0.003	0.01	3.63	0.10
05N 33E 17ADD1 USGS 28	298	434600112360101	USGS-INL	112°36'03"	43°45'59"	1,455.42	0.003	145	2	1,380.68	0.003	0.01	1.59	2.58
05N 33E 13BDC1 USGS 30C	299	434601112315401	USGS-INL	112°31'57"	43°46'00"	1,462.26	0.003	336	4	1,376.51	0.003	0.01	1.72	-0.68
05N 30E 11CDD1 DH 1B	300	434611112504301	USGS-INL	112°50'46"	43°46'10"	1,461.70	0.003	235	5	1,374.80	0.003	0.01	3.32	-1.43
05N 33E 10CDC1 USGS 31	301	434625112342101	USGS-INL	112°34'23"	43°46'25"	1,459.79	0.003	130	2	1,378.86	0.003	0.01	1.35	-0.98
05N 34E 09BDA1 USGS 4	302	434657112282201	USGS-INL	112°28'24"	43°46'55"	1,461.31	0.003	1,920	4	1,377.98	0.003	0.01	1.00	-5.52
06N 32E 36ADD1 2ND OWSLEY	303	434819112380501	USGS-INL	112°38'09"	43°48'19"	1,459.48	0.003	352	2	1,386.32	0.003	0.01	0.97	-0.12
06N 33E 26DDB1 USGS 27	304	434851112321801	USGS-INL	112°32'21"	43°48'50"	1,459.21	0.003	1,210	12	1,386.60	0.003	0.01	1.44	-9.79
06N 32E 26CDB1 ANP 9	305	434856112400001	USGS-INL	112°40'03"	43°48'55"	1,459.88	0.003	461	4	1,387.33	0.003	0.01	1.53	-0.06
06N 32E 26CAB1 ANP 10	306	434909112400401	USGS-INL	112°40'06"	43°49'08"	1,459.85	0.003	53	2	1,387.88	0.003	0.01	1.89	-0.23
06N 31E 27BDD1 USGS 7	307	434915112443901	USGS-INL	112°44'42"	43°49'14"	1,460.82	0.003	851	4	1,390.86	0.003	0.01	1.23	1.98
06N 31E 21DCC1 PSTF TEST	308	434941112454201	USGS-INL	112°45'44"	43°49'40"	1,459.94	0.003	91	3	1,390.18	0.003	0.01	2.41	0.33
06N 32E 22CCB2 GIN 3	309	434945112413101	USGS-INL	112°41'33"	43°49'44"	1,459.90	0.003	24	1	1,391.00	0.003	0.01	2.52	0.21

Table 5   65

Table 5. Wells in the Federal-State Cooperative and U.S. Geological Survey-Idaho National Laboratory water-level monitoring networks, eastern Snake River Plain, Idaho, during 2008.—Continued

Local name	Map No.	Site No.	Network name	Longitude	Latitude	Reference point elevation (m)	Reference point location error (m)	Sample size POR	Sample size 2008	2008 water-level elevation (m)	2008 measurement method error (m)	2008 measurement error (m)	Standard deviation (m)	Estimation error (m)
06N 31E 24DDA1 GIN 1	310	434947112414301	USGS-INL	112°41'43"	43°49'47"	1,460.05	0.003	22	1	1,390.82	0.003	0.01	2.39	-0.19
06N 32E 22CCB1 GIN 2	311	434949112413401	USGS-INL	112°41'36"	43°49'48"	1,459.90	0.003	41	1	1,391.02	0.003	0.01	2.39	-0.04
06N 32E 22CCB3 GIN 3	312	434949112413601	USGS-INL	112°41'37"	43°49'48"	1,459.93	0.003	22	1	1,391.05	0.003	0.01	2.39	0.05
06N 32E 22CBC1 GIN 5	313	434953112413301	USGS-INL	112°41'34"	43°49'53"	1,460.02	0.003	23	1	1,391.08	0.003	0.01	2.37	0.03
06N 32E 22BBD6 TAN 23A	314	435020112412704	USGS-INL	112°41'29"	43°50'19"	1,460.19	0.003	21	5	1,391.31	0.003	0.01	1.32	-0.05
06N 32E 22BBD1 TAN 15	315	435021112412701	USGS-INL	112°41'29"	43°50'20"	1,460.10	0.003	22	5	1,391.37	0.003	0.01	1.58	0.07
06N 31E 13CDD2 TAN 17	316	435034112421601	USGS-INL	112°42'18"	43°50'33"	1,460.90	0.003	68	6	1,391.13	0.003	0.01	1.52	-0.31
06N 31E 13CDD1 TAN 8	317	435034112421701	USGS-INL	112°42'19"	43°50'33"	1,461.52	0.003	68	6	1,391.44	0.003	0.01	1.44	0.32
06N 31E 16DCA1 NO NAME 1	318	435038112453401	USGS-INL	112°45'35"	43°50'38"	1,459.31	0.003	84	4	1,390.88	0.003	0.01	2.81	0.24
06N 31E 13CCA3 TAN 14	319	435039112423701	USGS-INL	112°42'39"	43°50'38"	1,458.27	0.030	23	5	1,390.94	0.003	0.03	1.55	-0.08
06N 31E 13CCA2 TAN 13A	320	435040112423801	USGS-INL	112°42'40"	43°50'39"	1,458.15	0.030	21	5	1,390.99	0.003	0.03	1.45	0.02
06N 31E 13DBB4 TAN 18	321	435051112421401	USGS-INL	112°42'17"	43°50'50"	1,464.99	0.003	22	5	1,391.46	0.003	0.01	1.30	-0.01
06N 31E 13DBB5 TAN 19	322	435051112421501	USGS-INL	112°42'17"	43°50'50"	1,465.12	0.003	21	5	1,391.52	0.003	0.01	1.31	0.07
06N 31E 13DBB1 USGS 24	323	435053112420801	USGS-INL	112°42'15"	43°50'50"	1,462.60	0.003	1,748	15	1,391.34	0.003	0.01	1.56	-0.10
06N 31E 11CDC1 FET DISP 3	324	435124112433701	USGS-INL	112°43'40"	43°51'23"	1,458.73	0.003	39	1	1,391.75	0.003	0.01	2.24	0.56
06N 31E 10ACC1 ANP 6	325	435152112443101	USGS-INL	112°44'34"	43°51'51"	1,462.40	0.003	88	2	1,391.16	0.003	0.01	2.44	-0.37
06N 31E 12ACD1 IET 1 DISP	326	435153112420501	USGS-INL	112°42'08"	43°51'53"	1,461.05	0.003	60	2	1,391.28	0.003	0.01	2.60	-0.21
06N 32E 11ABA1 USGS 26	327	435212112394001	USGS-INL	112°39'43"	43°52'10"	1,460.66	0.003	559	5	1,391.25	0.003	0.01	1.47	-10.83
07N 31E 33DCD1 ANP 5	328	435308112454101	USGS-INL	112°45'44"	43°53'07"	1,486.11	0.003	67	1	1,391.80	0.003	0.01	1.92	0.78

Table 5. Wells in the Federal-State Cooperative and U.S. Geological Survey-Idaho National Laboratory water-level monitoring networks, eastern Snake River Plain, Idaho, during 2008.—Continued

Local name	Map No.	Site No.	Network name	Longitude	Latitude	Reference point elevation (m)	Reference point location error (m)	Sample size POR	Sample size 2008	2008 water-level elevation (m)	2008 measurement method error (m)	2008 measurement error (m)	Standard deviation (m)	Estimation error (m)
07N 31E 34BDD1 USGS 25	329	435339112444601	USGS-INL	112°44'48"	43°53'38"	1,478.88	0.003	3,584	56	1,391.61	0.003	0.01	1.69	0.18
07N 31E 28CAC1 P and W 1	330	435416112460401	USGS-INL	112°46'06"	43°54'15"	1,493.26	0.003	69	1	1,391.85	0.003	0.01	1.88	0.92
07N 31E 28DAB1 P and W 2	331	435419112453101	USGS-INL	112°45'33"	43°54'18"	1,491.80	0.003	98	2	1,391.10	0.003	0.01	2.42	-0.44
07N 31E 26BBC1 P and W 3	332	435443112435801	USGS-INL	112°44'01"	43°54'42"	1,490.16	0.003	72	1	1,391.94	0.003	0.01	1.78	-1.32
07N 31E 22BDD1 ANP7	333	435522112444201	USGS-INL	112°44'46"	43°55'19"	1,505.15	0.003	60	2	1,391.81	0.003	0.01	1.91	-1.88
07N 31E 20BDB1 USGS 126A	334	435529112471301	USGS-INL	112°47'15"	43°55'28"	1,521.64	0.003	93	4	1,391.54	0.003	0.01	2.04	-0.00
07N 31E 20BDB2 USGS 126B	335	435529112471401	USGS-INL	112°47'16"	43°55'28"	1,521.81	0.003	93	4	1,391.52	0.003	0.01	2.03	-0.02

Table 6   67

Table 6. Wells identified for removal based on genetic algorithm searches, eastern Snake River Plain, Idaho.

[Number of sites removed: well sites removed from an existing monitoring network (n_r). Local name: local well identifier used in this study. Map No.: identifier used to locate wells on map figures and as a cross reference with data in other tables. Site No.: unique numerical identifiers used to access well data (http://waterdata.usgs.gov/nwis). Times identified: number of times the observation well was identified for removal in each of the networks five genetic algorithm searches. Control parameter values: a kriging grid resolution of 2.5 kilometers for the Co-op network and 1.5 kilometers for the USGS-INL network, population size of 2,000, elitism rate of 0.05, crossover probability of 0.80, mutation probability of 0.30, and terminates after 50 consecutive iterations without any improvement in the best fitness value. Abbreviations: Co-op network, Federal-State Cooperative water-level monitoring network; USGS-INL network, U.S. Geological Survey-Idaho National Laboratory water-level monitoring network]

Number of sites removed	Optimized Co-op network				Optimized USGS-INL network			
	Map No.	Local name	Site No.	Times identified	Map No.	Local name	Site No.	Times identified
10	8	11S 18E 25DDC1	422555114172101	5	184	02N 29E 18CCD2 RWMC M4D	432939113030101	5
	34	09S 29E 04BCA1	424013113043801	4	187	02N 29E 18DCB1 USGS 119	432945113023401	5
	80	04S 31E 20BBB1	430402112520301	5	188	02N 29E 18DCA1 USGS 118	432947113023001	5
	84	04S 33E 03CBB2	430601112353301	4	189	02N 29E 18CBD1 USGS 117	432955113025901	5
	91	03S 33E 14BBA1	431006112340901	4	206	03N 29E 36DDC1 ICPP 1798	433216112562601	4
	120	06N 39E 30ADC1	434915111540501	5	213	03N 29E 36BDB3 USGS 128	433250112565601	5
	124	06N 39E 28BBB1	434932111523701	5	245	03N 29E 24DDA3 USGS 46	433407112561501	5
	140	07N 36E 22ABD4	435528112121201	5	251	03N 29E 23CBA1 MIDDLE 1823	433418112581701	1
	146	07N 36E 09BBB1	435728112141301	4	314	06N 32E 22BBD6 TAN 23A	435020112412704	5
	164	09N 36E 04BAA1	440841112133001	5	321	06N 31E 13DBB4 TAN 18	435051112421401	5
20	8	11S 18E 25DDC1	422555114172101	5	184	02N 29E 18CCD2 RWMC M4D	432939113030101	5
	32	09S 26E 07AAB1	423943113272001	4	187	02N 29E 18DCB1 USGS 119	432945113023401	5
	33	09S 25E 03CAC1	424003113313101	4	188	02N 29E 18DCA1 USGS 118	432947113023001	5
	34	09S 29E 04BCA1	424013113043801	4	189	02N 29E 18CBD1 USGS 117	432955113025901	5
	72	05S 31E 19DDC2	425754112521601	4	190	02N 29E 18CBA1 RWMC M1SA	432956113030901	4
	80	04S 31E 20BBB1	430402112520301	5	194	02N 29E 18BDA1 USGS 87	433013113024201	4
	84	04S 33E 03CBB2	430601112353301	4	206	03N 29E 36DDC1 ICPP 1798	433216112562601	4
	91	03S 33E 14BBA1	431006112340901	4	213	03N 29E 36BDB3 USGS 128	433250112565601	5
	92	03S 34E 02BCC3	431126112271503	4	224	03N 29E 25CAD1 USGS 38	433322112564301	4
	120	06N 39E 30ADC1	434915111540501	5	230	03N 29E 25BDC1 USGS 34	433334112565501	4
	124	06N 39E 28BBB1	434932111523701	5	236	03N 29E 25AAA2 USGS 123	433352112561401	4
	131	06N 39E 10BBB1	435209111512101	2	237	03N 30E 30BBA2 USGS 122	433353112555201	4
	132	06N 38E 02DBD1	435228111563401	3	245	03N 29E 24DDA3 USGS 46	433407112561501	5
	137	07N 35E 20CBD1	435504112222301	4	248	03N 29E 24DAD1 USGS 40	433411112561101	4
	140	07N 36E 22ABD4	435528112121201	5	257	03N 29E 14DDA2 USGS 58	433500112572502	4
	146	07N 36E 09BBB1	435728112141301	4	266	03N 32E 14AAC1 ANL OBS A 001	433545112394101	4
	147	07N 34E 04CDC1	435728112281101	4	314	06N 32E 22BBD6 TAN 23A	435020112412704	5
	152	08N 36E 21DCD1	440002112131801	3	320	06N 31E 13CCA2 TAN 13A	435040112423801	4
	158	09N 36E 33CBB1	440353112135701	4	321	06N 31E 13DBB4 TAN 18	435051112421401	5
	164	09N 36E 04BAA1	440841112133001	5	322	06N 31E 13DBB5 TAN 19	435051112421501	4

Table 6. Wells identified for removal based on genetic algorithm searches, eastern Snake River Plain, Idaho.—Continued

Number of sites removed	Optimized Co-op network				Optimized USGS-INL network			
	Local name	Map No.	Site No.	Times identified	Local name	Map No.	Site No.	Times identified
40	12S 20E 04DBC1	3	4224241 14070001	3	02N 29E 18CCD2 RWMC M4D	184	4329391 13030101	5
	11S 18E 25DDC1	8	4225555 114172101	5	02N 29E 18DCB1 USGS 119	187	4329451 13023401	5
	10S 21E 26AAA2	16	4231591 13570302	3	02N 29E 18DCA1 USGS 118	188	4329471 13023001	5
	09S 26E 10DDD1	31	4238551 13233901	2	02N 29E 18CBD1 USGS 117	189	4329551 13025901	5
	09S 26E 07AAB1	32	4239431 13272001	4	02N 29E 18CBA1 RWMC M15A	190	4329561 13030901	4
	09S 25E 03CAC1	33	4240031 13313101	4	02N 29E 18BDA1 USGS 87	194	4330131 13024201	4
	09S 29E 04BCA1	34	4240131 13043801	4	03N 29E 36DDC1 ICPP 1798	206	4332161 12562601	4
	08S 27E 31DDA1	35	4240421 13201101	3	03N 29E 36BDB3 USGS 128	213	4332501 12565601	5
	08S 14E 16CBB1	47	4243531 14494701	3	03N 29E 25DDB1 USGS 113	218	4333141 12561801	3
	08S 27E 07DBC1	48	4244191 13201801	3	03N 29E 25DCA1 USGS 112	219	4333141 12563001	3
	05S 31E 19DDC2	72	4257541 12521601	4	03N 30E 30CCB1 USGS 77	220	4333151 12560301	3
	04S 31E 20BBB1	80	4304021 12520301	5	03N 30E 30CBD1 USGS 114	221	4333181 12555001	3
	04S 31E 11ABA1	81	4305471 12473701	3	03N 29E 25CAD1 USGS 38	224	4333221 12564301	4
	04S 33E 03CBB2	84	4306101 12353301	4	03N 29E 25BDD1 USGS 36	226	4333301 12565201	3
	03S 33E 14BBA1	91	4310061 12340901	4	03N 30E 30BCC1 USGS 111	228	4333311 12560501	3
	03S 34E 02BCC3	92	4311261 12271503	4	03N 29E 25BDC1 USGS 34	230	4333341 12565501	4
	02S 35E 11DDD1	96	4315171 12190101	3	03N 29E 25BDB1 USGS 35	231	4333391 12565801	3
	03N 37E 12BDB1	109	4336251 12031801	3	03N 29E 25BBD1 USGS 39	232	4333431 12570001	3
	06N 39E 35CBB2	118	4348161 11501302	3	03N 30E 30BAD1 USGS 67	233	4333441 12554101	3
	06N 39E 30ADC1	120	4349151 11540501	5	03N 29E 25ABD1 USGS 57	234	4333441 12562601	3
	06N 38E 25ACB4	121	4349171 11553102	2	03N 30E 30BBB1 USGS 51	235	4333501 12560601	3
	06N 37E 29ACA2	122	4349221 12072202	3	03N 29E 25AAA2 USGS 123	236	4333521 12561401	4
	06N 39E 28BBB1	124	4349321 11523701	5	03N 30E 30BBA2 USGS 122	237	4333531 12555201	4
	06N 39E 23AAC2	125	4350151 11495302	3	03N 30E 30BAB1 USGS 59	238	4333541 12554701	3
	06N 39E 16DAA1	127	4350481 11512701	3	03N 30E 19CCC1 USGS 48	241	4334011 12560301	3
	06N 39E 13ABA1	129	4351181 11481601	3	03N 29E 24DDC1 USGS 45	242	4334021 12561801	3
	07N 39E 34CCB1	133	4353141 11511902	3	03N 29E 24DDA2 USGS 42	243	4334041 12561301	3
	07N 35E 20CBD1	137	4355041 12222301	4	03N 30E 19CCB1 USGS 47	244	4334071 12560301	3
	07N 38E 23DBA2	138	4355061 11563102	3	03N 29E 24DDA3 USGS 46	245	4334071 12561501	5
	07N 40E 19ADD4	139	4355161 11464002	3	03N 29E 24DDA1 USGS 41	246	4334091 12561301	3
	07N 36E 22ABD4	140	4355281 12121201	5	03N 29E 24DAD1 USGS 40	248	4334111 12561101	4
	07N 39E 16DBB3	142	4356051 11515801	3	03N 29E 24DAD2 USGS 43	250	4334151 12561501	3
	07N 35E 13AAD1	143	4356261 12164301	3	03N 29E 14DDA2 USGS 58	257	4335001 12572502	4
	07N 36E 11ABB1	145	4357231 12111101	3	03N 32E 14AAC1 ANL OBS A 001	266	4335451 12394101	3
	07N 34E 04CDC1	147	4357281 12281101	4	04N 30E 29BCD1 NRF 11	278	4338471 12544201	3
	08N 41E 33ABB1	149	4359041 11373101	3	06N 32E 22BBD6 TAN 23A	314	4350201 12412704	5
	08N 34E 27CDD1	150	4359121 12264801	3	06N 31E 13CDD1 TAN 8	317	4350341 12421701	3
	08N 36E 21DCD1	152	4400021 12131801	3	06N 31E 13CCA2 TAN 13A	320	4350401 12423801	4
	09N 36E 33CBB1	158	4403531 12135701	4	06N 31E 13DBB4 TAN 18	321	4350511 12421401	5
	09N 36E 04BAA1	164	4408411 12133001	5	06N 31E 13DBB5 TAN 19	322	4350511 12421501	4

Table 6   69

Table 6. Wells identified for removal based on genetic algorithm searches, eastern Snake River Plain, Idaho.—Continued

Number of sites removed	Optimized Co-op network				Optimized USGS-INL network			
	Local name	Map No.	Site No.	Times identified	Local name	Map No.	Site No.	Times identified
60	12S 20E 04DBC1	3	422424114070001	3	02N 29E 18CCD2 RWMC M4D	184	432939113030101	5
	11S 22E 32CCC1	6	422501113543901	2	02N 29E 18DCB1 USGS 119	187	432945113023401	5
	11S 18E 25DDC1	8	422555114172101	5	02N 29E 18DCA1 USGS 118	188	432947113023001	5
	10S 21E 26AAA2	16	423159113570302	3	02N 29E 18CBD1 USGS 117	189	432955113025901	5
	09S 22E 33ADA1	21	423604113522401	2	02N 29E 18CBA1 RWMC MISA	190	432956113030901	4
	09S 16E 21DCD1	22	423722114345101	2	02N 29E 18BDA1 USGS 87	194	433013113024201	4
	09S 25E 23DBA1	23	423732113295801	2	03N 29E 36DDC1 ICPP 1798	206	433216112562601	4
	09S 26E 13CCC2	25	423802113222701	1	03N 29E 36DCC2 CFA LF 2-10	207	433216112563301	2
	09S 29E 18CDA1	26	423808113063601	2	03N 29E 36BCB1 USGS 85	212	433246112571201	2
	09S 26E 07AAB1	32	423943113272001	4	03N 29E 36BDB3 USGS 128	213	433250112565601	5
	09S 25E 03CAC1	33	424003113313101	4	03N 29E 25DDB1 USGS 113	218	433314112561801	3
	09S 29E 04BCA1	34	424013113043801	4	03N 29E 25DCA1 USGS 112	219	433314112563001	3
	08S 27E 31DDA1	35	424042113201101	3	03N 30E 30CCB1 USGS 77	220	433315112560301	3
	08S 25E 36DAA1	39	424102113282101	2	03N 30E 30CBD1 USGS 114	221	433318112555001	3
	08S 26E 33BCB2	41	424112113255402	2	03N 29E 25CAD1 USGS 38	224	433322112564301	4
	08S 14E 16CBB1	47	424353114449701	3	03N 29E 25BDD1 USGS 36	226	433330112565201	3
	08S 27E 07DBC1	48	424419113201801	3	03N 30E 30ACC1 USGS 116	227	433331112553201	2
	08S 14E 12CBC1	49	424439114461201	2	03N 30E 30BCC1 USGS 111	228	433331112560501	3
	08S 26E 03DCC1	50	424454113240101	2	03N 29E 25BDC1 USGS 34	230	433334112565501	4
	07S 25E 19BAA1	56	424828113345201	2	03N 29E 25BDB1 USGS 35	231	433339112565801	3
	05S 33E 35CDC1	68	425608112340901	2	03N 29E 25BBD1 USGS 39	232	433343112570001	3
	05S 31E 19DDC2	72	425754112521601	4	03N 30E 30BAD1 USGS 67	233	433344112554101	3
	05S 31E 27ABA1	73	425757112485201	2	03N 29E 25ABD1 USGS 57	234	433344112562601	3
	04S 31E 20BBB1	80	430402112520301	5	03N 30E 30BBB1 USGS 51	235	433350112560601	3
	04S 31E 11ABA1	81	430547112473701	3	03N 29E 25AAA2 USGS 123	236	433352112561401	4
	04S 32E 01CBA1	82	430607112400501	2	03N 30E 30BBA2 USGS 122	237	433353112555201	4
	04S 33E 03CBB2	84	430601112353301	4	03N 30E 30BAB1 USGS 59	238	433354112554701	3
	03S 33E 25CCC1	86	430729112331201	2	03N 29E 23DCD1 USGS 84	239	433356112574201	2
	03S 33E 17AAD1	90	430955112365001	2	03N 30E 19CCC1 USGS 48	241	433401112560301	3
	03S 34E 02BCC3	92	431126112271503	4	03N 29E 24DDC1 USGS 45	242	433402112561801	3
	02S 34E 33BBA1	94	431242112292801	2	03N 29E 24DDA2 USGS 42	243	433404112561301	3
	03S 35E 11DDD1	96	431517112190101	3	03N 30E 19CCB1 USGS 47	244	433407112560301	3
	03N 37E 12BDB1	109	433625112031801	3	03N 29E 24DDA3 USGS 46	245	433407112561501	5
	06N 39E 35CBB2	118	434816111501302	3	03N 29E 24DDA1 USGS 41	246	433409112561301	3
	06N 35E 27DDA1	119	434857112185801	2	03N 29E 24DAD1 USGS 40	248	433411112561101	4
	06N 39E 30ADC1	120	434915111540501	5	03N 29E 24DAD2 USGS 43	250	433415112561501	3
	06N 38E 25ACB4	121	434917111553102	2	03N 29E 23ADC1 USGS 76	253	433425112573201	2
	06N 37E 29ACA2	122	434922112072202	3	03N 29E 23ABB1 USGS 65	254	433447112574501	2
	06N 39E 28BBB1	124	434932111523701	5	03N 29E 14DDA2 USGS 58	257	433500112572502	4
	06N 39E 23AAC2	125	435015111495302	3	03N 29E 14CBD1 USGS 79	259	433505112581901	2

Table 6. Wells identified for removal based on genetic algorithm searches, eastern Snake River Plain, Idaho.—Continued

Number of sites removed	Optimized Co-op network				Optimized USGS-INL network			
	Local name	Map No.	Site No.	Times identified	Local name	Map No.	Site No.	Times identified
60—Continued	06N 35E 21AAB1	126	43502811202601	2	03N 29E 14ADD1 MTR TEST	262	43352011257260 1	2
	06N 39E 16DAA1	127	43504811151270 1	3	03N 32E 14AAC1 ANL OBS A 001	266	43354511239410 1	4
	06N 39E 13ABA1	129	435118111481601	3	04N 30E 30DAB1 NRF 9	275	43384011255020 1	2
	06N 36E 11ABA3	130	43520811210510 3	2	04N 30E 29CBB1 NRF 10	276	43384111254520 1	2
	06N 38E 02DBD1	132	43522811156340 1	3	04N 30E 30ADC1 NRF 8	277	43384311255090 1	2
	07N 39E 34CCB1	133	43531411151190 2	3	04N 30E 29BCD1 NRF 11	278	43384711254420 1	3
	07N 35E 20CBD1	137	43550411222230 1	4	04N 30E 19DDD1 NRF 6	281	43391011255010 1	2
	07N 38E 23DBA2	138	43550611156310 2	3	04N 30E 20CCA1 NRF 7	282	43392011254360 1	1
	07N 40E 19ADD4	139	43551611146400 2	3	06N 31E 24DDA1 GIN 1	310	43494711241430 1	2
	07N 36E 22ABD4	140	43552811212120 1	5	06N 32E 22CCB1 GIN 2	311	43494911241340 1	2
	07N 39E 16DBB3	142	43560511151580 1	3	06N 32E 22CCB3 GIN 4	312	43494911241360 1	2
	07N 35E 13AAD1	143	43562611216430 1	3	06N 32E 22CBC1 GIN 5	313	43495311241330 1	2
	07N 36E 11ABB1	145	43572311211110 1	3	06N 32E 22BBD6 TAN 23A	314	43502011241270 4	5
	07N 36E 09BBB1	146	43572811214130 1	4	06N 31E 13CDD1 TAN 8	317	43503411242170 1	3
	07N 34E 04CDC1	147	43572811228110 1	4	06N 31E 13CCA3 TAN 14	319	43503911242370 1	2
	08N 41E 33ABB1	149	43590411137310 1	3	06N 31E 13CCA2 TAN 13A	320	43504011242380 1	4
	08N 34E 27CDD1	150	43591211226480 1	3	06N 31E 13DBB4 TAN 18	321	43505111242140 1	5
	08N 36E 03DCD1	156	44023911212110 1	1	06N 31E 13DBB5 TAN 19	322	43505111242150 1	4
	09N 36E 33CBB1	158	44035311213570 1	4	06N 31E 13DDB1 USGS 24	323	43505311242080 1	2
	09N 36E 04BAA1	164	44084111213300 1	5	07N 31E 20BDB1 USGS 126A	334	43552911247130 1	2
80	12S 20E 04DBC1	3	42242411407000 1	3	02N 29E 19BCB1 USGS 120	180	43291911303150 1	1
	11S 22E 32CCC1	6	42250111354390 1	2	02N 29E 20BBA1 RWMC M6S	182	43293111301500 1	1
	11S 18E 25DDC1	8	42255511417210 1	5	02N 29E 18CCD2 RWMC M4D	184	43293911303010 1	5
	10S 21E 28BCB1	15	42314511400300 1	1	02N 29E 18CCD1 USGS 88	185	43294011303020 1	1
	10S 21E 26AAA2	16	42315911357030 2	3	02N 29E 18DCB1 USGS 119	187	43294511302340 1	5
	10S 17E 14CCD1	19	42325511426060 1	1	02N 29E 18DCA1 USGS 118	188	43294711302300 1	5
	09S 22E 33ADA1	21	42360411352240 1	2	02N 29E 18CBD1 USGS 117	189	43295511302590 1	5
	09S 16E 21DCD1	22	42372211434510 1	2	02N 29E 18CBA1 RWMC M1SA	190	43295611303090 1	4
	09S 25E 23DBA1	23	42373211329580 1	2	02N 29E 18BDA1 USGS 87	194	43301311302420 1	4
	09S 29E 18CDA1	26	42380811306360 1	2	02N 29E 17BBA1 RWMC M7S	196	43302311301480 1	1
	09S 28E 18BAD1	30	42383711313430 1	1	02N 29E 11CCA1 USGS 131	197	43303611258160 1	1
	09S 26E 10DDD1	31	42385511323390 1	2	02N 29E 09CDA2 USGS 129	198	43303611300270 1	1
	09S 26E 07AAB1	32	42394311327200 1	4	02N 29E 01DCA1 USGS 130	205	43313011256280 1	1
	09S 25E 03CAC1	33	42400311331310 1	4	03N 29E 36DCC2 CFA LF 2-10	207	43321611256330 1	2
	08S 27E 31DDA1	35	42404211320010 1	3	03N 29E 36CCC1 CFA LF 3-9	208	43321611257100 1	1
	08S 29E 34CBC1	36	42405211303390 1	1	03N 29E 36DAC1 CFA LF 2-11	211	43323011256170 1	1
	08S 24E 31DAC1	37	42405311341280 1	1	03N 29E 36BCB1 USGS 85	212	43324611257120 1	2
	08S 25E 36DAA1	39	42410211328210 1	2	03N 29E 36BDB3 USGS 128	213	43325011256560 1	5
	08S 26E 33BCB2	41	42411211325540 2	2	03N 29E 25DDB1 USGS 113	218	43331411256180 1	3

Table 6   71

Table 6. Wells identified for removal based on genetic algorithm searches, eastern Snake River Plain, Idaho.—Continued

Number of sites removed	Optimized Co-op network				Optimized USGS-INL network			
	Local name	Map No.	Site No.	Times identified	Local name	Map No.	Site No.	Times identified
80—Continued	08S 15E 32BBA1	42	424144114434101	1	03N 29E 25DCA1 USGS 112	219	433314112563001	3
	08S 14E 16CBB1	47	424353114494701	3	03N 30E 30CCB1 USGS 77	220	433315112560301	3
	08S 27E 07DBC1	48	424419113201801	3	03N 30E 30CBD1 USGS 114	221	433318112555001	3
	08S 14E 12CBC1	49	424439114461201	2	03N 30E 30CAD1 USGS 115	223	433320112554101	1
	08S 26E 03DCC1	50	424454113240101	2	03N 29E 25CAD1 USGS 38	224	433322112564301	4
	07S 25E 19BAA1	56	424828113345201	2	03N 29E 25BDD1 USGS 36	226	433330112565201	3
	07S 29E 12CCC2	57	424916113012001	1	03N 30E 30ACC1 USGS 116	227	433331112553201	2
	06S 34E 09BCB1	66	425456112294001	1	03N 30E 30BCC1 USGS 111	228	433331112560501	3
	05S 33E 35CDC1	68	425608112340901	2	03N 29E 26CAB1 ICPP MON A-167	229	433331112580701	1
	05S 31E 19DDC2	72	425754112521601	4	03N 29E 25BDC1 USGS 34	230	433334112565501	4
	05S 31E 27ABA1	73	425757112485201	2	03N 29E 25BDB1 USGS 35	231	433339112565801	3
	04S 33E 20CBB1	79	430333112375801	1	03N 29E 25BBD1 USGS 39	232	433343112570001	3
	04S 31E 20BBB1	80	430402112520301	5	03N 30E 30BAD1 USGS 67	233	433344112554101	3
	04S 31E 11ABA1	81	430547112473701	3	03N 29E 25ABD1 USGS 57	234	433344112562601	3
	04S 32E 01CBA1	82	430607112400501	2	03N 30E 30BBB1 USGS 51	235	433350112560601	3
	04S 31E 05CBC1	83	430607112515601	1	03N 29E 25AAA2 USGS 123	236	433352112561401	4
	03S 33E 25CCC1	86	430729112333201	2	03N 30E 30BBA2 USGS 122	237	433353112555201	4
	03S 33E 17AAD1	90	430955112365001	2	03N 30E 30BAB1 USGS 59	238	433354112554701	3
	03S 33E 14BBA1	91	431006112340901	4	03N 29E 23DCD1 USGS 84	239	433356112574201	2
	03S 34E 02BCC3	92	431126112271503	4	03N 30E 19DDC2 USGS 82	240	433401112551001	1
	02S 34E 33BBA1	94	431242112292801	2	03N 30E 19CCC1 USGS 48	241	433401112560301	3
	02S 35E 22DAC1	95	431349112202001	1	03N 29E 24DDC1 USGS 45	242	433402112561801	3
	02S 35E 11DDD1	96	431517112190101	3	03N 29E 24DDA2 USGS 42	243	433404112561301	3
	02S 33E 16ABB1	97	431520112360901	1	03N 30E 19CCB1 USGS 47	244	433407112560301	3
	03N 38E 22BAB1	108	433457111583701	1	03N 29E 24DDA3 USGS 46	245	433407112561501	5
	03N 37E 12BDB1	109	433625112031801	3	03N 29E 24DDA1 USGS 41	246	433409112561301	3
	05N 39E 08DAD1	115	434638111530401	1	03N 29E 24DDB1 USGS 44	247	433409112562101	1
	05N 36E 02BDA2	116	434748112113602	1	03N 29E 24DAD1 USGS 40	248	433411112561101	4
	06N 39E 35CBB2	118	434816111501302	3	03N 29E 24DAD2 USGS 43	250	433415112561501	3
	06N 35E 27DDA1	119	434857112185801	2	03N 29E 23ADC1 USGS 76	253	433425112573201	2
	06N 39E 30ADC1	120	434915111540501	5	03N 29E 23ABB1 USGS 65	254	433447112574501	2
	06N 37E 29ACA2	122	434922112072202	3	03N 29E 14DDA2 USGS 58	257	433500112572502	4
	06N 38E 30BAD2 USBR SITE 1	123	434924112013801	1	03N 29E 14CBD1 USGS 79	259	433505112581901	2
	06N 39E 28BBB1	124	434932111523701	5	03N 29E 14DBD1 TRA DISP	260	433506112572301	1
	06N 39E 23AAC2	125	435015111495302	3	03N 29E 14ADD1 MTR TEST	262	433520112572601	2
	06N 35E 21AAB1	126	435028112202601	2	03N 32E 14AAC1 ANL OBS A 001	266	433545112394101	4
	06N 39E 16DAA1	127	435048111512701	3	03N 29E 01DBB1 USGS 98	269	433657112563601	1
	06N 39E 13ABA1	129	435118111481601	3	04N 30E 30DAB1 NRF 9	275	433840112550201	2
	06N 36E 11ABA3	130	435208112105103	2	04N 30E 29CBB1 NRF 10	276	433841112545201	2

Table 6. Wells identified for removal based on genetic algorithm searches, eastern Snake River Plain, Idaho.—Continued

Number of sites removed	Optimized Co-op network				Optimized USGS-INL network			
	Local name	Map No.	Site No.	Times identified	Local name	Map No.	Site No.	Times identified
80—Continued	06N 39E 10BBB1	131	435209111512101	2	04N 30E 30ADC1 NRF 8	277	433843112550901	2
	06N 38E 02DBD1	132	435228111563401	3	04N 30E 29BCD1 NRF 11	278	433847112544201	3
	07N 39E 34CCB1	133	435314111511902	3	04N 30E 30ACA1 USGS 102	279	433853112551601	1
	07N 35E 20CBD1	137	435504112222301	4	04N 30E 29BAC1 NRF 12	280	433855112543201	1
	07N 38E 23DBA2	138	435506111563102	3	04N 30E 19DDD1 NRF 6	281	433910112550101	2
	07N 40E 19ADD4	139	435516111464002	3	04N 30E 19DAD1 NRF 13	283	433928112545401	1
	07N 36E 22ABD4	140	435528112121201	5	06N 32E 26CAB1 ANP 10	306	434909112400401	1
	07N 34E 24BBA1	141	435540112243901	1	06N 31E 24DDA1 GIN 1	310	434947112414301	2
	07N 39E 16DBB3	142	435605111515801	3	06N 32E 22CCB1 GIN 2	311	434949112413401	2
	07N 35E 13AAD1	143	435626112164301	3	06N 32E 22CCB3 GIN 4	312	434949112413601	2
	07N 36E 11ABB1	145	435723112111101	3	06N 32E 22CBC1 GIN 5	313	434953112413301	5
	07N 36E 09BBB1	146	435728112141301	4	06N 32E 22BBD6 TAN 23A	314	435020112412704	1
	07N 34E 04CDC1	147	435728112281101	4	06N 32E 22BBD1 TAN 15	315	435021112412701	3
	08N 41E 33ABB1	149	435904111373101	3	06N 31E 13CDD1 TAN 8	317	435034112421701	2
	08N 34E 27CDD1	150	435912112264801	3	06N 31E 13CCA3 TAN 14	319	435039112423701	4
	08N 36E 21DCD1	152	440002112131801	3	06N 31E 13CCA2 TAN 13A	320	435040112423801	5
	08N 34E 11DCC1	154	440151112252301	1	06N 31E 13DBB4 TAN 18	321	435051112421401	4
	08N 40E 06CCC1	155	440261112474701	1	06N 31E 13DBB5 TAN 19	322	435051112421501	2
	09N 36E 33CBB1	158	440353112135701	4	06N 31E 13DBB1 USGS 24	323	435053112420801	1
	09N 34E 29DAB1	159	440447112284401	1	06N 31E 11CDC1 FET DISP 3	324	435124112433701	1
	09N 36E 15CCC1	160	440608112125001	1	07N 31E 28DAB1 P AND W 2	331	435419112453101	1
	09N 36E 04BAA1	164	440841112133001	5	07N 31E 20BDB1 USGS 126A	334	435529112471301	2

Appendix A. R-Package Documentation

Appendix A is available for viewing or download at http://pubs.usgs.gov/sir/2013/5120/.

Appendix B. Information about the R Session

Appendix B is available for viewing or download at http://pubs.usgs.gov/sir/2013/5120/.

www.ingramcontent.com/pod-product-compliance
Lightning Source LLC
Chambersburg PA
CBHW081558170526
45166CB00009B/2739